AF795192

AI and Gamification Technologies for Complex Work

The medium through which training in the workplace is delivered has been changing in recent years to offer a more personalized and immersive experience. The invention of virtual reality (VR) and augmented reality (AR) platforms has created opportunities to take a more hands-on approach to familiarizing oneself with a task or environment with mitigated time and monetary commitments. Written assessments are being swiftly replaced with more interactive and scientifically validated training simulations and this essential technology is in high demand in the government and private sectors. This book highlights many of the ways simulation-based training can be leveraged to create personalized training curricula for those in high-risk careers and how it can be assessed successfully.

AI and Gamification Technologies for Complex Work uncovers the use of artificial intelligence (AI) and machine learning (ML) for the purposes of creating adaptive, personalized training for individuals who work in complex jobs. It covers adaptive simulation-based training, fighting skill decay through game-based training, and additional uses of AI/ML and other tools in measuring human performance. Insights from professionals and experts in the fields of simulation and training provide readers with information about current applications of AI/ML in creating adaptive or personalized training, as well as investigations into the future of simulation and game-based training, as virtual and augmented realities proliferate modern training programs. The book looks at how data science, AI, and ML contribute to adaptive training systems and the reader is encouraged to look further into the engines that drive adaptive training while devising their own systems for training in complex jobs.

This book is ideal for professionals in human factors engineering and psychology, artificial intelligence, military training and simulation, game development, data science, modeling and simulation and industrial and organizational psychology.

AI and Gamification Technologies for Complex Work

Simulation and Training

Edited by
Phillip M. Mangos
and James C. Ferraro

CRC Press
Taylor & Francis Group
Boca Raton London New York

CRC Press is an imprint of the
Taylor & Francis Group, an **informa** business

Designed cover image: Shutterstock

First edition published 2026
by CRC Press
2385 NW Executive Center Drive, Suite 320, Boca Raton FL 33431

and by CRC Press
4 Park Square, Milton Park, Abingdon, Oxon, OX14 4RN

CRC Press is an imprint of Taylor & Francis Group, LLC

© 2026 selection and editorial matter, Phillip M. Mangos and James C. Ferraro; individual chapters, the contributors

Reasonable efforts have been made to publish reliable data and information, but the author and publisher cannot assume responsibility for the validity of all materials or the consequences of their use. The authors and publishers have attempted to trace the copyright holders of all material reproduced in this publication and apologize to copyright holders if permission to publish in this form has not been obtained. If any copyright material has not been acknowledged please write and let us know so we may rectify in any future reprint.

Except as permitted under U.S. Copyright Law, no part of this book may be reprinted, reproduced, transmitted, or utilized in any form by any electronic, mechanical, or other means, now known or hereafter invented, including photocopying, microfilming, and recording, or in any information storage or retrieval system, without written permission from the publishers.

For permission to photocopy or use material electronically from this work, access www.copyright.com or contact the Copyright Clearance Center, Inc. (CCC), 222 Rosewood Drive, Danvers, MA 01923, 978-750-8400. For works that are not available on CCC please contact mpkbookspermissions@tandf.co.uk

Trademark notice: Product or corporate names may be trademarks or registered trademarks and are used only for identification and explanation without intent to infringe.

ISBN: 978-1-032-65076-0 (hbk)
ISBN: 978-1-032-70161-5 (pbk)
ISBN: 978-1-032-70163-9 (ebk)

DOI: 10.1201/9781032701639

Typeset in Times
by KnowledgeWorks Global Ltd.

Contents

Foreword

INTRODUCTION

Contemporary workplaces are increasingly characterized by complexity, automation, and the need for sophisticated training and assessment techniques to keep pace with technological advancement. Within this evolving landscape, artificial intelligence (AI) and gamification have emerged as transformative technologies for enhancing human performance across various domains. From military training to healthcare simulation, and from human resource processes to team development, organizations are seeking innovative methods to bridge the gap between current capabilities and future needs. This edited volume represents a comprehensive exploration of how AI and gamification are reshaping performance assessment, training methodologies, and decision-making processes in complex work environments. The authors, drawing from diverse backgrounds in psychology, computer science, military training, healthcare, human factors, and industrial/organizational psychology, provide evidence-based insights into the theoretical foundations, practical applications, and ethical considerations of these emerging technologies. By examining the intersection of AI, gamification, and complex work, this collection offers a timely and valuable resource for researchers, practitioners, and leaders interested in leveraging cutting-edge technologies to enhance human performance and organizational effectiveness.

UNIFYING THEMES

Three primary themes emerge across the chapters in this volume: adaptive training technologies, human-AI collaboration, and the ethical dimensions of AI implementation. These themes serve as conceptual bridges connecting the various domains and applications discussed throughout the book.

Adaptive Training Technologies

A central theme throughout this volume is the development and implementation of adaptive training technologies that respond dynamically to individual learner needs. Adaptive training systems, powered by increasingly sophisticated AI algorithms and frameworks, represent a significant advancement over traditional "one-size-fits-all" approaches to training and education. These systems continuously monitor learner performance, assess comprehension and skill acquisition in real time, and adjust instructional content, difficulty levels, and feedback mechanisms accordingly. The fundamental premise underlying adaptive training is that personalization enhances learning outcomes by ensuring that instruction consistently targets each learner's "sweet spot" – challenging enough to promote growth while avoiding boredom, frustration, and content with limited learning value. As demonstrated in various chapters, these adaptive systems serve diverse training contexts, from military close air support missions to healthcare simulations to team coordination exercises. The empirical evidence presented consistently indicates that well-designed adaptive training systems can significantly improve learning efficiency, enhance skill transfer, and boost learner engagement. What makes these systems particularly powerful is their ability to capture and process multimodal data – including text, speech, physiological signals, and behavioral patterns – to develop increasingly robust and accurate models of learner competence and performance. By leveraging these rich data sources, adaptive training technologies can provide more precise instruction and more meaningful feedback than traditional training approaches, ultimately accelerating skill development and improving performance outcomes in complex operational environments.

HUMAN-AI COLLABORATION

The second unifying theme explores the evolving relationship between humans and AI in complex work environments, emphasizing collaboration rather than replacement. Across chapters, the authors consistently highlight that the most effective implementations of AI and gamification technologies are those that augment human capabilities rather than attempt to automate them entirely. This perspective represents a significant shift from earlier concerns about AI displacing human workers toward a more nuanced understanding of the complementary strengths of human and artificial intelligence. In personnel selection and assessment contexts, AI systems can analyze vast quantities of candidate data, identify patterns, and predict performance outcomes with impressive accuracy; yet, human judgment remains essential for contextualizing these insights and making final decisions. In training environments, AI can adapt content and provide immediate feedback, but human instructors are crucial for establishing learning objectives, designing meaningful scenarios, and facilitating reflective practice. Even in highly automated domains, the research presented demonstrates that human oversight, intervention capabilities, and strategic decision-making remain indispensable elements of effective human-machine systems. The concept of human-AI collaboration extends beyond mere division of labor to encompass trust building, explainability, and interface design considerations that optimize interaction between human users and AI systems. By focusing on collaboration rather than replacement, this volume offers a more productive and realistic vision of AI implementation that acknowledges both the remarkable capabilities of artificial intelligence and the unique contributions of human judgment, creativity, and adaptability.

ETHICAL DIMENSIONS OF AI IMPLEMENTATION

The third unifying theme addresses the ethical considerations and challenges associated with implementing AI in human performance assessment and training. As AI systems become increasingly integral to decision-making processes that affect human careers, development opportunities, and work experiences, ensuring fairness, transparency, and accountability becomes paramount. Several chapters highlight concerns regarding algorithm bias, noting that AI systems trained on historical data may perpetuate or even amplify existing inequities in selection, evaluation, and promotion decisions. Various authors emphasize the importance of rigorous validation processes, ongoing monitoring, and diverse training data to mitigate algorithmic bias. Privacy considerations represent another significant ethical dimension, as AI-enhanced training and assessment systems typically collect and analyze extensive personal data, from interview responses to physiological reactions to behavioral patterns. The responsible governance of this data, including clear policies regarding collection, storage, usage, and access rights, features prominently in discussions of ethical implementation. Finally, the issue of transparency runs through multiple chapters, with authors advocating for explainable AI (XAI) approaches that enable stakeholders to understand how and why specific recommendations or decisions are reached. By confronting these ethical dimensions directly, the volume provides a balanced perspective that acknowledges both the tremendous potential of AI technologies and the responsibility to implement them in ways that respect human dignity, promote fairness, and earn stakeholder trust. This nuanced treatment of ethical considerations offers valuable guidance for researchers and practitioners seeking to harness the benefits of AI while avoiding pitfalls that could undermine its acceptance and effectiveness.

About the Editors

Phillip M. Mangos, PhD, is the CEO and Chief Scientist of Adaptive Immersion Technologies, a Florida-based business focused on the synthesis of predictive data modeling and analytics, simulation, and assessment technology to optimize human performance. He holds a BS in Psychology from the University of South Florida and a PhD in Industrial/Organizational and Human Factors from Wright State University. With a diverse career background, he has worked as a team member to support projects in aviation, law enforcement, transportation, intelligence, information technology, and utilities industries.

James C. Ferraro, PhD, is a Senior Human Factors Research Scientist at Adaptive Immersion Technologies. He specializes in intelligent simulations and game-based assessments to improve human performance in complex, automated systems. He holds a PhD in Human Factors and Cognitive Psychology and an MA in Applied Experimental and Human Factors Psychology from the University of Central Florida. His research focuses on human-machine interaction, trust in automation, and performance prediction. Dr. Ferraro has contributed to government-sponsored projects and co-edited multiple book series on human performance and simulation, with numerous publications in the field.

Contributors

Alejandro Arca
Adaptive Immersion
Tampa, FL

Shannon K. T. Bailey
Center for Advanced Medical Learning and Simulation
University of South Florida Health
Tampa, FL

Gautam Biswas
Department of Computer Science
Institute for Software Integrated Systems
Vanderbilt University
Nashville, TN

Marc Cubrich
APT*Metrics*
Chicago, IL

Crystal M. Fausett
San Jose State University
San Jose, CA

James C. Ferraro
Adaptive Immersion
Tampa, FL

Carter Gibson
HireVue
Kalamazoo, MI

Benjamin Goldberg
U.S. Army DEVCOM – Soldier Center
Simulation and Training Technology Center (STTC)
Orlando, FL

Gabriella M. Hancock
California State University
Long Beach, Long Beach, CA

P. A. Hancock
University of Central Florida
Orlando, FL

Jason E. Hochreiter
Naval Air Warfare Center Training Systems Division
Orlando, FL

Ian M. Hughes
Department of Psychological and Brain Sciences
Texas A&M University
College Station, TX

Cheryl I. Johnson
Leidos
Orlando, FL

Theresa T. Kessler
Human Centered Engineering Division
Georgia Tech Research Institute
Atlanta, GA

Rachel T. King
Vero AI
Cleveland, OH

John Licato
Computer Science and Engineering Department
University of South Florida
Tampa, FL

Phillip M. Mangos
Adaptive Immersion
Tampa, FL

Matthew D. Marraffino
Naval Air Warfare Center Training Systems Division
Orlando, FL

Cory Moore
HireVue
Fort Collins, CO

Laura M. Ornelas
California State University
Long Beach, Long Beach, CA

Andrew Samo
Department of Psychology,
Bowling Green State University
Bowling Green, OH

Tracy L. Sanders
MITRE Corporation
McLean, VA

Bradford L. Schroeder
Naval Air Warfare Center Training
Systems Division
Orlando, FL

Richard J. Simonson
Department of Quality, Safety,
and Human Factors
Children's Mercy Hospital
Kansas City, MO
Department of Pediatrics
University of Missouri-Kansas City,
School of Medicine
Kansas City, MO

Wendi L. Van Buskirk
Naval Air Warfare Center Training
Systems Division
Orlando, FL

Caleb Vatral
Department of Computer Science
Tennessee State University
Nashville, TN

1 A Theoretical Framework for Performance Analysis in Competency-Based Experiential Learning Environments

Caleb Vatral, Gautam Biswas, and Benjamin Goldberg

INTRODUCTION

The successful integration of complex cognitive and psychomotor skills in both individual and team contexts is necessary for competent performance outcomes in today's workplaces. Due to this growing complexity and demand, effective learning and training programs must be developed that teach these abilities in realistic yet safe contexts. Competency-based experiential learning environments are an increasingly popular paradigm for this kind of training and learning, where students are trained in real-world situations through hands-on interaction with the subject. As experiential learning environments continue to proliferate, there is an increasing need for robust modeling schemes capable of effectively assessing both individual and group learning performances and behaviors. Nevertheless, the implementation of efficient and effective learner models in these complex environments poses several challenges. These challenges encompass the accurate tracking and interpretation of learners' psychomotor activities and cognitive behaviors in both space and time, the precise monitoring of their progress across the multiple psychomotor and cognitive dimensions, and the generation of adaptive personalized feedback that is constructive and addresses individual learners' difficulties and suboptimal performance by providing personalized recommendations.

In this chapter, we present a theoretical framework for multimodal learner modeling, performance analysis, and feedback generation to support debriefing and after-action reviews (AARs). Our framework, which is grounded in the theories of competency-based education (CBE) and experiential learning, combines multimodal learning analytics (MMLA), distributed cognition, and cognitive task analysis (CTA) to produce an effective model of learner performance and behaviors and generate feedback interventions. Our framework utilizes a mixed-methods approach that systematically breaks down a training task into its constituent components,

DOI: 10.1201/9781032701639-1

develops measures and algorithms to support task assessments, and then generates understandable and actionable feedback for the instructor and the trainees that helps to explain the generated assessments.

As a first step, our *CTA* methodology generates a comprehensive hierarchical model of the relevant cognitive, behavioral, and psychomotor tasks and problem-solving strategies employed by the learners. Using the CTA model as a guide, we conduct a *distributed cognition analysis* to interpret and categorize learner data captured across multiple modalities (such as visual, speech, physiological, and logged activities in the mixed-reality (MR) environment). This analysis aids in generating needs analysis and feature engineering to support MMLA, effectively organizing and interpreting raw learner data. We leverage strong foundations in CBE to construct a hierarchical assessment model that is populated from our analyses of the multimodal data. This hierarchical learner competency model utilizes *Bayesian inferencing* techniques to aggregate information and generate insights about learners' cognitive states at multiple levels of abstraction. The information from the hierarchical task model can be mapped onto *performance metrics* that can be calculated across time. Finally, using the insights generated by the hierarchical assessment model, we generate *feedback* that is designed to be presented back to learners as well as their instructors to help them understand how the assessment metrics were generated. In addition, we can provide suggestions for *actionable information* on how to improve learners' performance.

CASE STUDY

In the chapter, we will use a case study of soldiers training in dismounted battle drills in MR environments to illustrate each component of our learner modeling framework (Figure 1.1). Then, using the lessons learned from this case study, we will discuss the application of our framework to a broader class of experiential learning and training environments.

In Fall 2021, infantry fire teams participated in a study where they trained on two dismounted battle drills at Fort Campbell US Army Installation: Enter and Clear a Room (ECR) and Break Contact (BC). The ECR drill involves entering a new room, neutralizing enemy combatants, securing civilians, and exiting safely. The BC drill involves exploring a region with potential enemy forces, breaking contact,

FIGURE 1.1 The two dismounted battle drills run on the SAM-T and used for this case study. Left: Break Contact; Right: Enter and Clear a Room.

and retreating to a safe distance. Both drills require proper protocols and best practices to minimize risks and casualties. The Squad Advanced Marksmanship Trainer (SAM-T) system, an MR battle drill simulator, was used for training. The SAM-T projects a Virtual Battle Space 3 (VBS3) simulation onto screens setup in a physical environment space, allowing the team to move around in the physical space while simultaneously interacting with the simulation using modified weapons. The system reacts adaptively to soldier weapon fire and instructors could also modify the simulation in real time in response to other trainee actions. The system recorded log information about simulation events, including weapon fire messages and entity positions. Video and audio of the trainees were collected using the Generalized Intelligent Framework for Tutoring (GIFT). The data was synchronized with the VBS3 and SAM-T logs, allowing for offline processing by our proposed methods. The recorded data and subsequent analyses can be played back through a GIFT Gamemaster interface to support debriefing and AAR.

BACKGROUND

Competency-Based Education

CBE is a learner-centric approach that focuses on mastering specific competencies and skills, rather than traditional curriculum (Carraccio et al., 2002). It allows for personalized learning paths and flexible pacing, aligning with modern workplace demands and focusing on the attainment of relevant skill sets for organizational success. CBE, rooted in behaviorist theories of education, gained its first wave of advocacy in the late 1960s through the 1980s (Morcke et al., 2013). However, it faced increased scrutiny in the mid-1970s due to the shift away from the behavioral objectives curriculum model. The third wave of advocacy began in the early 2010s, with influential publications, renewed federal policy, and critical review of empirical evidence (Cooke et al., 2010; Gervais, 2016; Johnstone & Soares, 2014). Critics of CBE argue that its behaviorist foundation links curriculum to assessment and regulation of proficiency, rather than teaching and learning activities (Naranjo, 2022). They contend that, in addition to learning behaviorally, students learn affectively and socially, which cannot be objectively specified for competency goals (Stenhouse, 1975). Later arguments in response to this humanistic criticism have tried to expand CBE to harmonize with a constructivist curriculum framework (Morcke et al., 2013). In this work, we adopt this constructivist approach by harmonizing elements of CBE with the humanistic and constructivist *experiential learning* theory, thus only using CBE as part of a larger framework of CBEL, as will be described in the next sections.

One of the fundamental aspects of CBE programs is that competencies must be measurable, using various assessment methods like exams, rubric-based demonstrations, self-assessment reflections, and competency portfolios (Rowan, 2016). Best practices suggest measuring competency across the course curriculum in a variety of ways to establish program validity and comprehensively evaluate students. Classical competency assessment methods, while beneficial for education, also pose challenges due to the need for significant human judgment. Without a comprehensive standardized evaluation protocol, evaluations and grades can be labor-intensive for instructors and incomparable across classes (Allen, 2005; York et al., 2015). In this work, we work toward solving these significant challenges by constructing a comprehensive theoretical framework for collection, analysis, and presentation of multimodal learner activity data in competency-based experiential training environments.

Experiential Learning

Experiential learning is a pedagogical method that encourages active engagement, reflection, and practical application of knowledge and skills. It bridges the gap between theory and practice, allowing learners to construct knowledge in real-world contexts. Kolb's Theory of Experiential Learning, rooted in Dewey, Levin, and Piaget's experiential work, is a significant theoretical framework for experiential learning (Kolb, 2014; McCarthy, 2017). It posits that learning is a dynamic process, involving direct experiences, and consists of four iterative and adaptive stages.

1. **Concrete Experience:** Refers to the initial encounter or experience with something new or unfamiliar, involving a learner actively engaging in the experience, either through direct observation or through participation.
2. **Reflective Observation:** After the concrete experience, individuals reflect on the experience and observe what happened, paying attention to the feelings, thoughts, and reactions they had during the experience.
3. **Abstract Conceptualization:** In this stage, learners attempt to make sense of the experience by creating theories or generalizations. They seek to understand the underlying principles or concepts that explain the observed events.
4. **Active Experimentation:** The final stage involves applying the theories and concepts derived from the reflective observation and conceptualization to a new situation or task. It is the stage of testing and applying the newly acquired knowledge and skills.

Kolb's Theory of Experiential Learning emphasizes active participation, reflection, and knowledge application in the learning cycle, making it a valuable framework for designing impactful experiences. While the theory is not without some criticism, it remains highly influential in experiential learning. Our work adopts Kolb's Theory of Experiential Learning as the primary theoretical basis for the design of the framework. Among the most common criticisms of experiential learning, there is still a great deal of ambiguity regarding the design of the *experiences* that make up each component of Kolb's cycle (Morris, 2020). This issue has been present since the early days of experiential learning, with scholars like John Dewey recognizing that not all experiences are equally educative. More recently, a systematic review of experiential learning literature has noted a lack of consensus regarding the definition of an *experience* among practitioners of Kolb's model, which serves to further substantiate earlier criticisms of Kolb's model, which describe it as "highly muddled" in that regard (Bergsteiner et al., 2010). In this work, we propose that this problem of how to design experiences can be resolved by harmonizing the field of experiential learning with the field of CBE.

Merging of Two Fields: Modern Research in CBEL

The integration of CBE and experiential learning has been a topic of research for many years, dating back to the publication of Reynolds in 1981 (Reynolds, 1981). Scholars have found a significant practical union between these two educational paradigms, with some even suggesting they are explicitly dependent on each other (Keenan, 2013). This has led to the term *Competency-Based Experiential Learning* being coined to describe the joint field. Building on this more recent work (Hoessler & Godden, 2021; Owens & Goldberg, 2022), in this chapter, we argue that the harmonized field of CBEL solves two major theoretical and methodological challenges associated with CBE and experiential learning alone.

First, combining CBE with experiential learning fosters a humanistic constructivist approach, distinguishing CBEL from the behaviorist origins of CBE. This combined approach specifies relational and soft-skill competencies often missing in CBE models. For example, Hoessler & Godden (2021) specifies many relational and intangible competency categories in their listing of CBEL/OBEL outcomes, including interpersonal qualities, growth and integration, student and society relations, etc. Since Kolb's model explicitly involves learners' humanistic reflections on their feelings about their experiences, this approach allows for effective development and evaluation of these competencies. Second, combining CBE with experiential learning allows for more grounded, observable experiences, addressing the design challenge of concrete experiences. This approach aligns competency outcomes, activities, and assessments around a common understanding of the experience's vision and purpose. Deeper learning is supported by learning experiences that have constructively matched objectives, activities, and assessments (Biggs & Tang, 2007).

While there are clearly significant benefits to the combined field of CBEL, there also remain significant challenges, including the challenges of effective assessment and feedback, which this chapter's proposed framework is designed to address. Assessment and feedback are crucial for effective learning, but there are challenges related to personalization and scalability of formative feedback mechanisms. Proper personalization involves frequent, specific feedback for each learner, which benefits learners by developing unique skills. However, this requires significant human judgment from instructors and can lead to incomparable evaluations and grades. Additionally, more frequent and specific feedback is more effective but requires more time for each student. To combat these issues surrounding personalization and scalability, this chapter proposes a comprehensive framework for automated assessment and feedback generation within CBEL environments. By building AI-driven computational tools to support these processes, we can support instructors in delivering timely, frequent, and specific feedback to a large number of students of various populations. In the next section, we will detail the design of the proposed framework at a conceptual level, which will then be followed by a discussion of an example implementation.

THEORETICAL FRAMEWORK

Our theoretical framework can be broadly characterized by the Input-Process-Output (IPO) structure (Figure 1.2). We use multimodal learner data collected during a training exercise as the first component of the model's input. Depending on

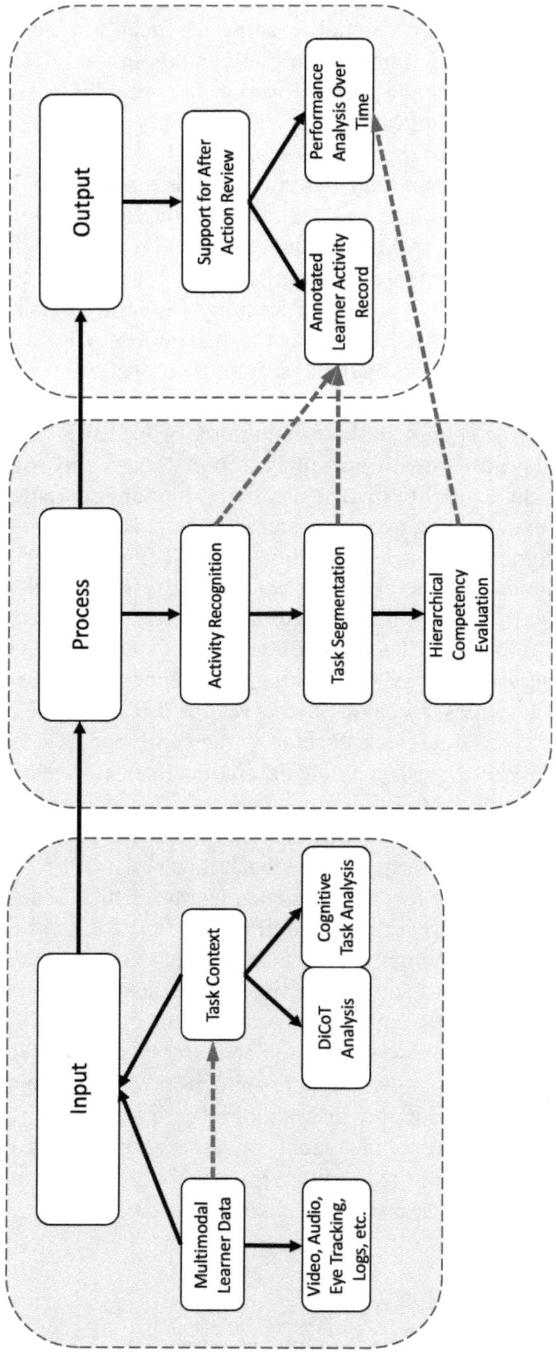

FIGURE 1.2 The high-level structure of the theoretical framework for learner modeling, evaluation, and feedback generation.

the particular training setting, this data may take many different forms, although it frequently consists of recording audio, video, system logs, and eye tracking data. Then, the task context, the system's other input, is built using a qualitative analysis of this data. Specifically, we perform CTA and DiCoT analysis, which will be covered in more detail in the subsequent sections. We include this task context as an input into the model as every task environment has distinct qualities that call for a corresponding unique analysis. Next, in the processing step, the model uses a three-phase *MMLA* algorithm to analyze the raw learner data through the lens of the task context. In order to transform the unstructured learner data into an organized and interpretable format, we first conduct *action recognition*. Next, we carry out *task segmentation*, which divides the entire training exercise into distinct sections that can be thoroughly examined, based on the recognized actions. Third, the hierarchical competency evaluation phase receives each of these chunks and uses the learner data to create a series of evaluations based on the expected behaviors specified in the task context. Finally, the output phase receives the processed learner data and related evaluations and converts them into feedback that the instructors and trainees may use for debriefing and AAR procedures. In our implementation of the framework, this learner feedback takes two forms: (1) an *annotated learner activity record* that allows instructors and trainees to quickly and easily navigate and review a training session and (2) *performance analysis* over time, which allows instructors and trainees to easily monitor learner progress and areas for improvement.

Cognitive Task Analysis

At the center of the analysis framework is the cognitive task model, which represents the primary learner model of the system. CTA is a methodical framework for understanding the cognitive mechanisms behind complex tasks performed by individuals and teams (Schraagen et al., 2000). It involves analyzing and breaking down tasks to identify underlying cognitive activities, decision-making processes, and problem-solving techniques. CTA builds a hierarchical model, representing cognitive processes involved in task execution. High levels represent abstract cognitive constructs, while each deepening level represents concrete and observable learner behaviors. The model's hierarchical structure allows for inferences about higher-level cognitive constructs, behaviors, strategies, and plans by understanding sequences of lower-level observable learner behavior (Biswas et al., 2019). This approach links low-level observable actions to higher-level strategies and behaviors. CTA models are created through a structured qualitative analysis of information from multiple sources. A preliminary analysis identifies the task's goals, subtasks, and cognitive demands. Domain experts and task performers provide additional insights through interviews, observations, and think-aloud protocols. These data sources are synthesized to create hierarchical structures that explain the sequential flow of cognitive activities during task execution. These models provide a comprehensive understanding of the cognitive complexities involved in task performance.

This work builds upon a fairly substantial history of the use of CTA for learner modeling. Much of this historical work has been focused on applications to K12 open-ended learning environments. Previous research has combined hierarchical CTA with sequencing mining to understand learners' strategies in the Betty's Brain teachable agent learning environment (Kinnebrew et al., 2017). Similar techniques have been applied to CTSiM and C2STEM learning environments to generate adaptive scaffolding and understand student problem-solving strategies (Emara et al., 2021). Outside of the K12 domain, CTA methods have also been applied for modeling learner behavior in adult training environments domains. In medical training domains, studies have found that CTA methods can elicit tacit knowledge and improve clinical practice (Swaby et al., 2022). Military training has also seen the application of CTA methods, with simulation-based training for medical command (Cannon-Bowers et al., 2013) and hierarchical CTA applied to military counterterrorism exercises like UrbanSim (Biswas et al., 2019). In our prior work, we utilized CTA to generate a hierarchical model and associated quantitative metrics for the ECR and BC dismounted battle drills (Vatral et al., 2023a, 2023b).

Figure 1.3 shows a partial cognitive task model for teamwork within our battle drill case study domain. From this model, it is easy to see the hierarchical

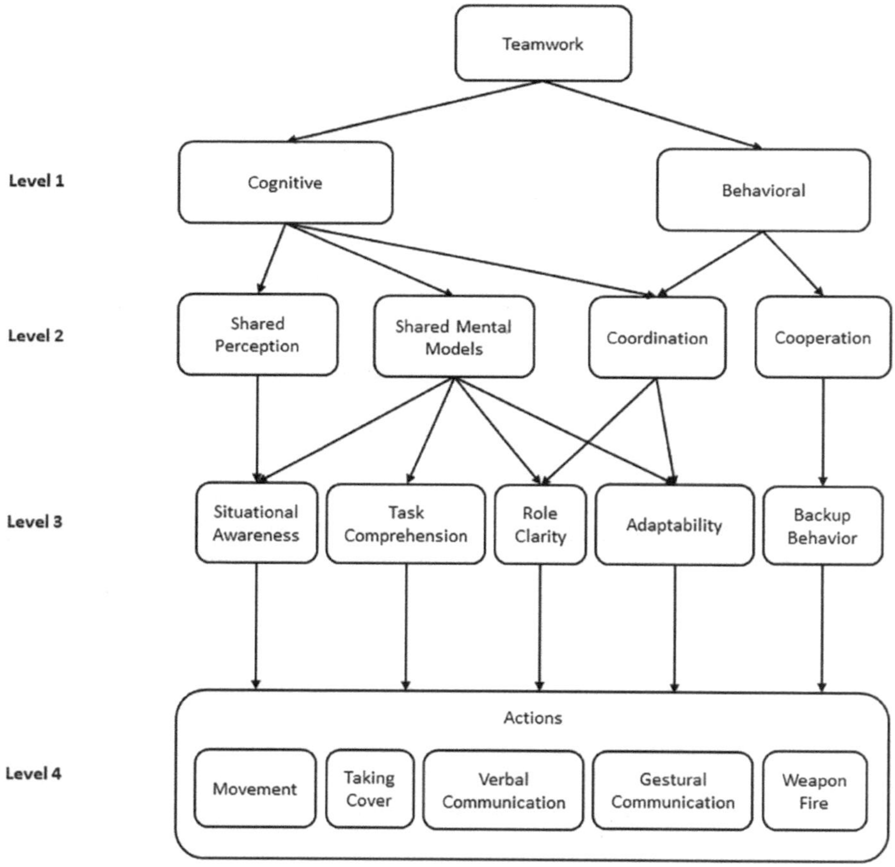

FIGURE 1.3 Example cognitive task model for teamwork in the Army battle drill case study domain.

structure of the tasks, breaking down each high-level idea into further subtasks and eventually into observable actions. These lowest-level actions are observed and sequenced to infer the higher-level task a participant is performing. For example, in the BC drill, if we observe a soldier quickly change from movement to taking cover, we might infer that the soldier has changed higher-level tasks due to contact with enemy forces. Based on these higher-level actions and the transitions between them, we can then perform *Event Segmentation*, which divides the entire training scenario into smaller parts that can be independently examined by the competency model (Zhang et al., 2021). We will discuss this in greater detail later. Because every higher-level action in the task model may have wildly different criteria and, thus, be scored wildly differently, event segmentation is a crucial stage in the assessment process.

However, the cognitive task model by itself is unable to fully contextualize higher-level actions in complicated and open-ended training contexts. Considering the BC drill's movement patterns from our case study, without additional information, we are unable to determine if the team's basic movement is a component of the break phase that follows contact with the enemy troops, or if it is part of the initial exploration phase. Instead, we need to provide more domain information in order to clarify which phase this movement falls in. In our framework, this additional contextual information is provided by a distributed cognition analysis.

Distributed Cognition

Our theoretical framework extends the insights generated by CTA by combining the CTA model with additional domain information provided by a distributed cognition analysis. Distributed cognition is a theory of human cognition that challenges the traditional individualistic view of cognition by suggesting that cognition occurs not only in the mind of the individual alone, but rather as a complex interaction between individual minds, other people, and the environment in which the cognition is taking place (Hutchins, 2000). Distributed cognition views the complete activity system as the unit of study rather than the individual mind, with the aim of comprehending cognition at this system level (Hazlehurst et al., 2008; Rybing, 2018). According to Hutchins, cognition occurs in three modalities: within a social group, between internal and external structures, and across time (Hutchins, 2000). Social group members collaborate to solve problems and contribute to a common goal. Internal structures, such as tools, offload cognitive processing, while physical space layout defines the affordances available to participants. Time and temporal evolution is also important, as earlier events can influence the nature of later events.

The application of distributed cognition requires the selection of one of several methodological frameworks that implements the basic theory. Each of these frameworks has been developed to study distributed cognition applied to various domains and scenarios. Examples include the Resource Model for human computer interaction (Wright et al., 2000), the Determining Information Flow Breakdown model for organizational learning in medical settings (Galliers et al., 2007), and the Event Analysis of Systemic Teamwork framework for submarine interactions (Stanton, 2014). Because of the importance of teamwork in CBEL environments, as well as following its wide adoption in analyzing trainee behaviors in this work, we adopt the *Distributed Cognition for Teamwork (DiCoT)* model proposed by Blandford and Furniss (2006).

DiCoT is a qualitative analysis framework that categorizes a cognitive system into five themes: information flow, artifact and environment, physical layout, social interactions, and temporal evolution. It focuses on how information flows, how tools aid cognition, how objects and people are arranged, how social interactions affect cognition, and how the system changes over time. The DiCoT methodology defines 18 principles to analyze themes and their interactions in a distributed cognitive system. For example, principle 10: Information Hubs describes that certain artifacts in the system are central focuses where different channels of information meet. This principle is primarily related to the *information flow* and *artifacts and environment* themes. By analyzing the distributed cognition system and identifying the manifestations of all 18 principles, we can understand how each DiCoT theme contributes to the overall cognition.

Within the presented framework, we operationalize this DiCoT analysis computationally through the use of probabilistic constraints. Using BC as an example again, our DiCoT study may have shown that the enemy danger is displayed on the screens in the training environment. Thus, soldiers moving away from the screen would be considered a retreat and would map onto the break phase. This additional information allows us to infer that, in our case study, movement toward the SAM-T screen indicates that it is occurring during the exploration phase, and movement away from the screen indicates that it is occurring during the break phase. The task model, which lacks a comprehensive interpretation of the evolving scenario, alone cannot interpret the specific phase of the exercise that the troops are engaged in, unless supplemented with additional domain-specific information from the DiCoT model. In addition, the probabilistic constraints resulting from qualitative DiCoT analysis provide a methodical way to use a priori domain information to enhance the overall learner model.

MULTIMODAL LEARNING ANALYTICS

Action Recognition

Action recognition, in the context of the proposed framework, is the process of converting raw sensor data into interpretable learner actions. The process's specifics depend on the sensors available in the task environment and the actions that need to be recognized. In our previous work, we used computer vision algorithms for recognition of these actions, as video data was available for all of the recorded activities, but the methods presented here are designed to be general, allowing for the application of various analysis algorithms based on available data and desired outcomes. Design of the action recognition algorithms should be based on a combination of the available sensor data, the lowest observable levels of the cognitive task model, and the probabilistic constraints from the DiCoT analysis. For example, in the BC domain, we have previously demonstrated the importance of movement patterns in both the CTA and DiCoT analysis (Vatral et al., 2023a). In previous work, we have employed computer vision-based motion tracking algorithms to convert the raw video data into meaningful soldier position and movement data (Vatral et al., 2022, 2023a).

Hierarchical Competency Modeling

Next in the proposed framework is the hierarchical competency model, which represents the primary assessment step of the complete process. The hierarchical competency model is a structured model for providing an assessment of learner competency at multiple levels of abstraction. The structure of the competency model mirrors that of the cognitive task model, sharing many of its high-level components. Since these concepts are already domain-general in the CTA model, they require little transformation to convert them to parallel transferable competencies. However, instead of having observable behaviors at the lowest level of the model, as in CTA, the hierarchical competency model has computable assessment metrics at the lowest level. The metrics, crafted through the previously described qualitative analysis as well as consultation with domain experts, are designed to be computable from the directly observable learner data and the associated action recognition schemes from the previous steps. For example, when evaluating the BC drill, one metric might evaluate whether the trainees are staying close to cover when they are not moving. Example hierarchical competency models for the two dismounted battle drills in our case study are shown in Figure 1.4.

The parallel structure between the competency model and the cognitive task model allows us to take the event segmentation generated by the task model and generate the relevant assessment metrics for each segment depending on what task is being performed. These assessments for each segment are then propagated up the model to generate assessments of higher-level performance. To perform the propagation, we model the competency model as a dynamic Bayesian network, with directly computable low-level metrics representing the evidence variables, and higher-level competencies representing the unobservable variables conditionally linked to the evidence variables (Ben-Gal, 2008; Vatral et al., 2022).

We represent each node in the HCM as a variable in the Bayes net. These variables are characterized by one of three values – below-expectation, at-expectation, or above-expectation. This classification aligns with the three-state learner models utilized in both the GIFT (Goldberg et al., 2021), where our system is implemented, and the broader training literature (Cassella, 2010; Klein & Hoffman, 1992; Sottilare et al., 2017). Each competency node is assigned a prior probability distribution, reflecting the initial likelihood of a trainee or team being in a specific state for that competency. This mechanism enables the encapsulation of the initial experience level, with low experience indicated by a higher probability of below-expectation and, conversely, high experience indicated by a higher probability of above-expectation. For each link in the conceptual competency model, we establish a corresponding mathematical link in the Bayes net through conditional probability distributions. These distributions encode the relationship between two competencies. The performance on a higher-level transferable competency is thus conditionally dependent on demonstrating proficiency in low-level domain-specific competencies. This setup allows us to infer the states of unobservable competencies based on the evidence of low-level measurement competencies and metrics using Bayesian inference.

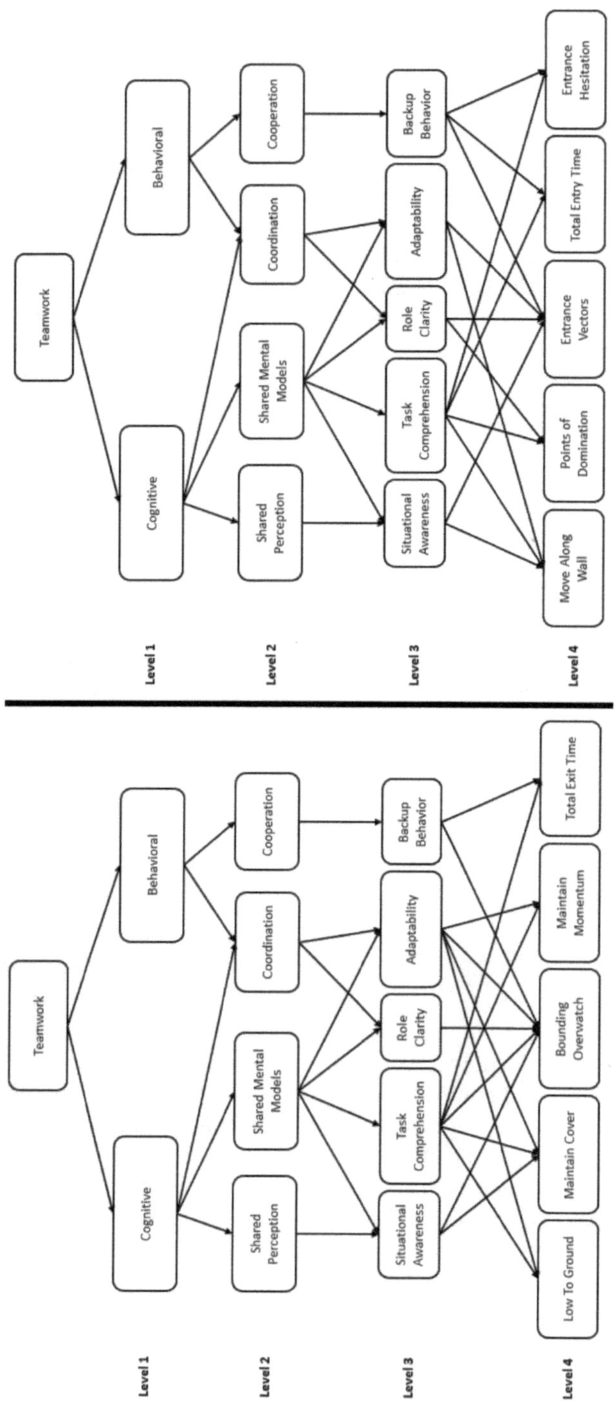

FIGURE 1.4 Example hierarchical competency models with connected domain-specific metrics for the Enter and Clear a Room (Left) and Break Contact (Right) dismounted battle drills.

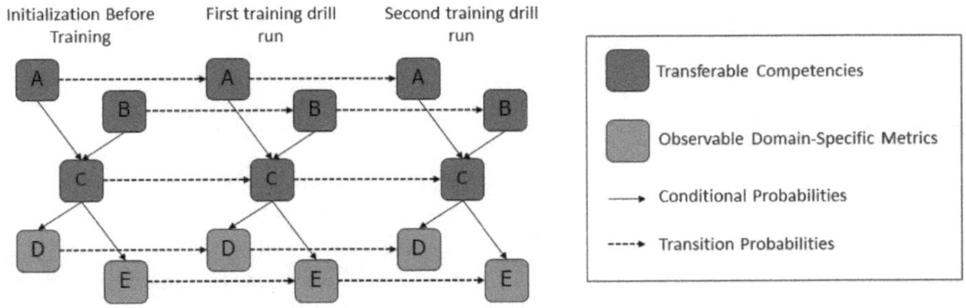

FIGURE 1.5 Illustration of the dynamic Bayes network update procedure.

Figure 1.5 provides an illustration of this process by displaying the links among unobservable transferable competences, observable domain-specific metrics, and updates across time. First, the prior-probability distributions are used to initialize the competency model. Subsequently, domain-specific metrics obtained from a training exercise serve as evidence variables. Using Bayesian inference on the conditional probabilities and prior probabilities, the evidence at these domain-specific nodes at time t is utilized to infer the states of the transferable competences during that training exercise. Then, the newly calculated states are used along with the transition probabilities to calculate the prior probabilities for inference at time $t + 1$ and the process repeats itself. Details of the probabilities used in this calculation for the case study are presented in the next sections.

In our case study, we utilize a set of manually created conditional and transition probability models, as shown in Table 1.1, and initialize the prior probability at time $t = 0$ to 100% below-expectation. The hand-designed conditional probability model was designed with the general idea that, since mastery of one concept depends on mastery of the other, parent and child concepts in the H-ABC hierarchy are very likely in the same state. When a child has more than one parent, the conditional probability distribution of all of the parent nodes is multiplied to get the complete conditional probability distribution of the child. The general concept of the hand-designed transition model is that, although it is highly unlikely that a trainee will move from one competency state to another following a single training event, the

TABLE 1.1
The Hand-Designed Probability Models Used in the H-ABC Bayesian Network for Our Case Study

	(a) Conditional Model				(b) Transition Model		
	Below	**At**	**Above**		**Below**	**At**	**Above**
Below	0.75	0.2	0.05	**Below**	0.95	0.05	0
At	0.2	0.6	0.2	**At**	0	0.95	0.05
Above	0.05	0.2	0.75	**Above**	0	0	1

likelihood of doing so increases if numerous training instances are completed consecutively, as is the case in our case study.

Feedback Generation

Last, the output provided to the end user is the final element of the theoretical framework. This research, at a broad level, aims to improve learning outcomes for students and trainees in experiential learning environments. Thus, providing feedback to learners and instructors is of critical importance. Feedback is a crucial tool for learners to improve their performance, reduce errors, and enhance their self-efficacy and motivation. It offers guidance and information in response to performance or understanding, enabling learners to refine their educational goals, scaffold strategies, and improve their self-efficacy (Adarkwah, 2021; Hattie & Timperley, 2007; Tan et al., 2020). Effective feedback has been shown to significantly impact learning outcomes, with effect sizes in systematic analyses ranging from moderate, $d = 0.48$, to moderately high, $d = 0.79$ (Hattie & Timperley, 2007; Wisniewski et al., 2020). However, not all feedback is equally effective, as it comes in various forms from various sources.

AAR, also known as debriefing, is perhaps the most widely used approach to provide feedback in experiential learning and CBEL domains. AAR involves a combination of feedback, reflection, and discussion following a training event. This process is a key component of Kolb's cycle of experiential learning and has been widely adopted in experiential learning and CBEL domains (Abulebda et al., 2022; Keiser & Arthur Jr., 2021). AAR techniques date back to the mid-1970s, popularized by the US Army. Since then, AAR-style methods have gained popularity in military applications and other training domains, including healthcare (Villado & Arthur Jr., 2013). Despite some variability in their development and implementation, the success of AAR is well documented across various domains and implementations, with meta-analyses reporting effect sizes of at least $d = 0.67$ and up to $d = 0.92$ (Keiser & Arthur Jr., 2022; Tannenbaum & Cerasoli, 2013).

The basic structure of an AAR is a three-phase process: (1) Reaction/Description, (2) Understanding/Analysis, and (3) Application/Summary (Abulebda et al., 2022). In the reaction/description phase, trainees are given the opportunity to diffuse and decompress from the training event, focusing on their feelings and the basic facts of events. This phase is considered important for psychological safety and prepares trainees for the rest of the debriefing (Twigg, 2020). In the understanding/analysis phase, instructors and trainees discuss the details of the training event, comparing what happened to what should have happened in ideal circumstances. Evidence suggests that discussions should focus on both successes of the trainees and areas for improvement. The inclusion of objective performance records, such as video and audio recordings, also improves the efficacy of the AAR (Keiser & Arthur Jr., 2021; Villado & Arthur Jr., 2013). In the application/summary phase, discussions focus on generalizing knowledge and applying lessons from the current learning experience to the future. This phase of AAR mirrors the final phase of Kolb's experiential learning model, allowing trainees to generalize what they have learned and improve their longitudinal performance.

Under the proposed framework of this chapter, the feedback generated by our algorithmic techniques is designed to primarily support the understanding/analysis phase of AAR. Data dashboards are used to present the generated feedback to stakeholders in a visual and interpretable way. The concept is to provide users with data and insights produced by the system in an annotated format so they can quickly review recorded performance data, understand how the underlying algorithms generate their specific recommendations, and understand how the generated feedback may be helpful in enhancing performance. While each implementation of the feedback mechanisms in the proposed framework will differ depending on the specific learning environment and its goals, three fundamental principles remain consistent in the design of each system: (1) feedback should be objective and data-driven, (2) users should understand how and why feedback was generated, and (3) feedback should supplement and assist traditional instructors, not replace them.

In our case study, we facilitate the feedback generation using the Gamemaster interface in the GIFT, whether our case study system was implemented (Goldberg et al., 2021). The Gamemaster in GIFT is an interface that allows users to review their performance on both current and previous training tasks. In our work, we have expanded the functionality of the Gamemaster interface to provide the user with two additional feedback components. First, we extended Gamemaster to display additional performance metrics displayed side-by-side with video that helps to demonstrate how these metrics were computed. This sort of annotated video timeline not only allows instructors and trainees to quickly move around the captured video to review specific segments corresponding to areas for improvement, but also allows for an in-depth review of annotated video evidence to understand how the algorithmic assessments come to their conclusions. Second, we expanded the Gamemaster interface to show a longitudinal graphical representation of trainee performance. Since the evaluations in the framework use a multilevel competency model, we can plot the performance of the trainees at each level of the model longitudinally to help give students and instructors an understanding of how they are progressing at multiple levels of abstraction. An example of this longitudinal performance plot is shown in Figure 1.6. In previous studies, we have found that these hierarchical performance visualizations can reveal significant insights into learner performance, including identifying concrete areas for improvement, illustrating task learning saturation to identify when training exercises should be modified, and examining how skills are transferring between multiple similar exercises (Vatral et al., 2022, 2023a).

CONCLUSIONS

In this study, we proposed a comprehensive theoretical framework that makes use of multimodal data from competency-based experiential training environments to analyze learner performance at various levels of abstraction. The framework offers a complete analytical methodology to take raw learner data and construct computational models that provide learner competence assessments and learner feedback. These models are grounded in strong theoretical foundations of CBEL, CTA, DiCoT, and MMLA. Throughout the chapter, we demonstrated the framework using a real-world scenario of fire teams of soldiers performing dismounted battle drill exercises.

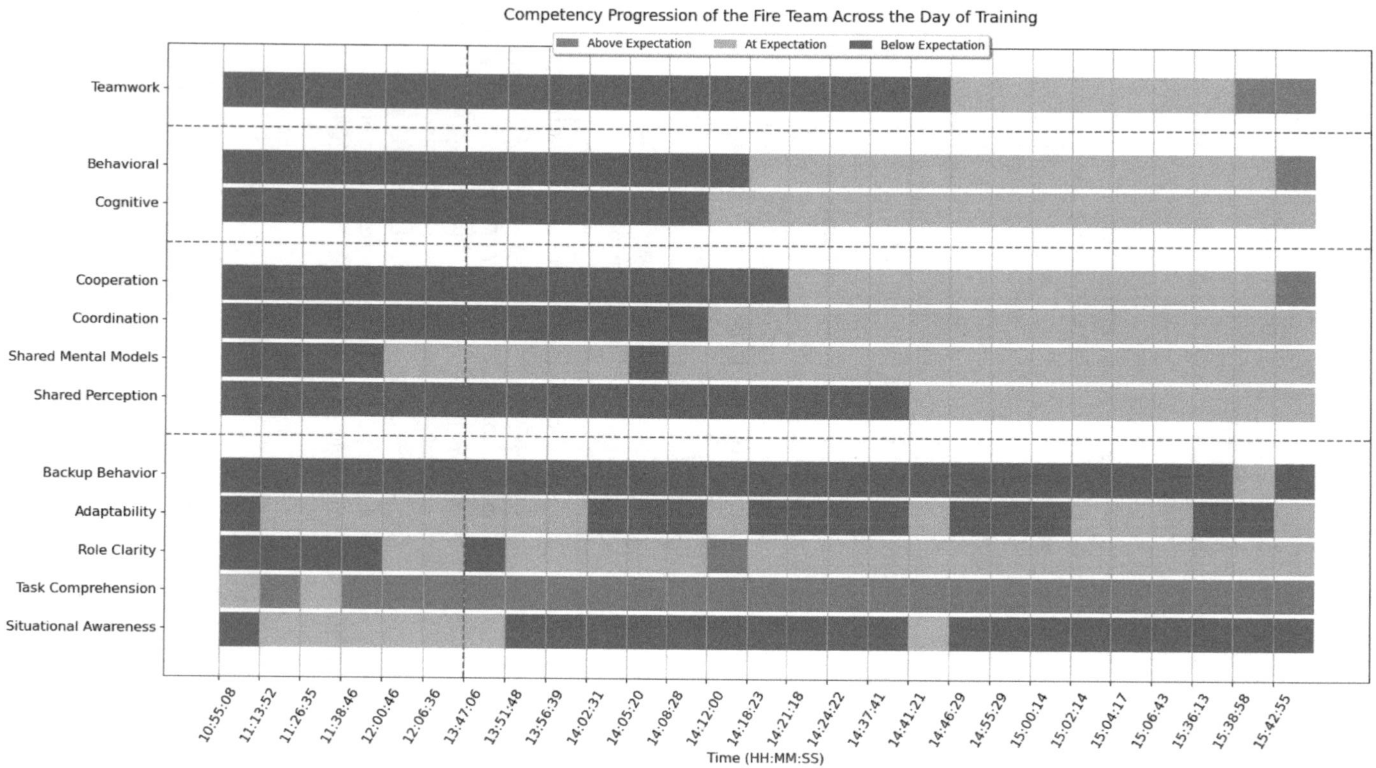

FIGURE 1.6 Visualization of performance progression across each level of the hierarchical competency model over the course of the entire day of training.

Future work should focus on applying this same computational framework to other cases besides Army battle drills, which would further validate the framework's efficacy. In addition, continuing work will focus on building out open-source extensible libraries that implement various algorithms of this framework that would be used by every implementation. These open-source implementations would allow for easy widespread adoption of the framework in a variety of training domains. In addition, by using a common toolset, these implementations could become interoperable with one another, potentially allowing for the construction of a more comprehensive learner model that builds from data from several unique training exercises. We anticipate that, with further development, this framework will serve as a thorough analytical procedure for a broad range of training areas and be crucial to the standardization of evaluation and feedback in competency-based experiential learning settings.

REFERENCES

Abulebda, K., Auerbach, M., & Limaiem, F. (2022). *Debriefing Techniques Utilized in Medical Simulation.* Treasure Island, FL: StatPearls.

Adarkwah, M. A. (2021). The power of assessment feedback in teaching and learning: A narrative review and synthesis of the literature. *SN Social Sciences, 1*(3).

Allen, J. (2005). Grades as valid measures of academic achievement of classroom learning. *The Clearing House: A Journal of Educational Strategies, Issues and Ideas, 78*(5), 218–223.

Ben-Gal, I. (2008). Bayesian Networks. In F. Ruggeri, R. S. Kenett, & F. W. Faltin (Eds.), *Encyclopedia of Statistics in Quality and Reliability.* Hoboken, NJ, USA: John Wiley & Sons Ltd. https://doi.org/10.1002/9780470061572.eqr089

Bergsteiner, H., Avery, G., & Neumann, R. (2010). Kolb's experiential learning model: Critique from a modeling perspective. *Studies in Continuing Education, 32*(1), 29–46.

Biggs, J., & Tang, C. (2007). *Teaching for Quality Learning at University.* Open University Press.

Biswas, G., Rajendran, R., Mohammed, N., Goldberg, B. S., Sottilare, R. A., Brawner, K., & Hoffman, M. (2019). Multilevel learner modeling in training environments for complex decision making. *IEEE Transactions on Learning Technologies, 13*(1), 172–185.

Blandford, A., & Furniss, D. (2006). DiCoT: A Methodology for Applying Distributed Cognition to the Design of Teamworking Systems. In *Interactive Systems. Design, Specification, and Verification* (pp. 26–38). Berlin, Heidelberg: Springer.

Cannon-Bowers, J., Bowers, C., Stout, R., Ricci, K., & Hildabrand, A. (2013). Using cognitive task analysis to develop simulation-based training for medical tasks. *Military Medicine, 178*, 15–21.

Carraccio, C., Wolfsthal, S., Englander, R., Ferentz, K., & Martin, C. (2002). Shifting paradigms: From Flexner to competencies. *Academic Medicine, 77*(5), 361–367.

Cassella, R. A. (2010). Leader development by design. *ITEA Journal, 31*, 280–283.

Cooke, M., Irby, D., & O'Brien, B. (2010). *Educating Physicians: A Call for Reform of Medical School and Residency* (Vol. *16*). San Francisco, CA: John Wiley & Sons.

Emara, M., Hutchins, N. M., Grover, S., Snyder, C., & Biswas, G. (2021). Examining student regulation of collaborative, computational, problem-solving processes in open-ended learning environments. *Journal of Learning Analytics, 8*(1), 49–74.

Galliers, J., Wilson, S., & Fone, J. (2007). A method for determining information flow breakdown in clinical systems. *International Journal of Medical Informatics, 76*, S113–S121.

Gervais, J. (2016). The operational definition of competency-based education. *The Journal of Competency-Based Education, 1*(2), 98–106.

Goldberg, B., Owens, K., Hellman, K., Robson, R., Blake-Plock, S., Hoffman, M., & Gupton, K. (2021). Forging Competency and Proficiency through the Synthetic Training Environment with an Experiential Learning for Readiness Strategy. In *Proceedings of the 2021 I/ITSEC*. Orlando, FL.

Hattie, J., & Timperley, H. (2007). The power of feedback. *Review of Educational Research, 77*(1), 81–112.

Hazlehurst, B., Gorman, P., & McMullen, C. (2008). Distributed cognition: An alternative model of cognition for medical informatics. *International Journal of Medical Informatics, 77*(4), 226–234.

Hoessler, C., & Godden, L. (2021). *Outcome-Based Experiential Learning: Let's Talk About, Design For, and Inform Teaching, Learning, and Career Development*. Higher Education and Beyond.

Hutchins, E. (2000). Distributed Cognition. In J. D. Wright (Ed.), *International Encyclopedia of the Social and Behavioral Sciences* (p. 138). Amsterdam: Elsevier Science.

Johnstone, S., & Soares, L. (2014). Principles for developing competency-based education programs. *Change: The Magazine of Higher Learning, 46*(2), 12–19.

Keenan, D. (2013). Experiential learning and outcome-based education: A bridge too far within the current education and training paradigm. *Journal of Applied Learning Technology, 3*(2), 13–19.

Keiser, N., & Arthur Jr., W. (2021). A meta-analysis of the effectiveness of the after-action review (or debrief) and factors that influence its effectiveness. *Journal of Applied Psychology, 106*(7), 1007–1032.

Keiser, N., & Arthur Jr., W. (2022). A meta-analysis of task and training characteristics that contribute to or attenuate the effectiveness of the after-action review (or debrief). *Journal of Business and Psychology, 37*(5), 953–976.

Kinnebrew, J. S., Segedy, J. R., & Biswas, G. (2017). Integrating model-driven and data-driven techniques for analyzing learning behaviors in open-ended learning environments. *IEEE Transactions on Learning Technologies, 10*(2), 140–153.

Klein, G. A., & Hoffman, R. R. (1992). Seeing the Invisible: Perceptual-Cognitive Aspects of Expertise. In M. Rabinowitz (Ed.), *Cognitive Science Foundations of Instruction* (pp. 203–226). Mahwah, NJ: Erlbaum.

Kolb, D. A. (2014). *Experiential Learning: Experience as the Source of Learning and Development*. Englewood Cliffs, NJ: FT Press.

McCarthy, J. (2017). Enhancing feedback in higher education: Students' attitudes towards online and in-class formative assessment feedback models. *Active Learning in Higher Education, 18*(2), 127–141.

Morcke, A. M., Dornan, T., & Eika, B. (2013). Outcome (competency) based education: An exploration of its origins, theoretical basis, and empirical evidence. *Advances in Health Sciences Education, 18*, 851–863.

Morris, T. H. (2020). Experiential learning–a systematic review and revision of Kolb's model. *Interactive Learning Environments, 28*(8), 1064–1077.

Naranjo, N. R. (2022). Criticisms of the Competency-Based Education (CBE) Approach. In Opačić A (ed), *Social Work in the Frame of a Professional Competencies Approach* (pp. 21–35). Cham: Springer International Publishing.

Owens, K., & Goldberg, B. (2022). Competency-based experiential expertise. In *Design Recommendations for Intelligent Tutoring Systems*, Volume 9-Competency-Based Scenario Design, 19.

Reynolds, C. R. (1981). Neuropsychological assessment and the habilitation of learning: Considerations in the search for the aptitude x treatment interaction. *School Psychology Review, 10*(3), 343–349.

Rowan, B. (2016). Defining Competencies and Outlining Assessment Strategies for Competency Based Education Programs. Pearson Education.

Rybing, J. (2018). *Studying Simulations with Distributed Cognition* (Vol. *1913*) Linkoping University Electronic Press.

Schraagen, J. M., Chipman, S. F., & Shalin, V. L. (Eds.). (2000). *Cognitive Task Analysis* (392 pp). Mahwah, NJ: Psychology Press.

Sottilare, R. A., Brawner, K. W., Sinatra, A. M., & Johnston, J. H. (2017). An updated concept for a generalized intelligent framework for tutoring (GIFT). GIFTtutoring. *org*, 1–19.

Stanton, N. (2014). Representing distributed cognition in complex systems: How a submarine returns to periscope depth. *Ergonomics*, *57*(3), 403–418.

Stenhouse, L. (1975). *An Introduction to Curriculum Research and Development.* Heinemann Publishers, London, UK.

Swaby, L., Shu, P., Hind, D., & Sutherland, K. (2022). The use of cognitive task analysis in clinical and health services research — A systematic review. *Pilot and Feasibility Studies*, *8*(1), 57.

Tan, F., Whipp, P., Gagne, M., & Van Quaquebeke, N. (2020). Expert teacher perceptions of two-way feedback interaction. *Teaching and Teacher Education*, *87*, 102930.

Tannenbaum, S., & Cerasoli, C. (2013). Do team and individual debriefs enhance performance? A meta-analysis. *Human Factors*, *55*(1), 231–245.

Twigg, S. (2020). Clinical event debriefing: A review of approaches and objectives. *Current Opinion in Pediatrics*, *32*(3), 337–342.

Vatral, C., Biswas, G., & Goldberg, B. (2023b). A theoretical framework for multimodal learner modeling and performance analysis in experiential learning environments. *Workshop on Artificial Intelligence in Support of Guided Experiential Learning*, Held in conjunction with the International Conference on Artificial Intelligence in Education (AIED). 07 July 2023, Tokyo, Japan (CEUR-WS.org).

Vatral, C., Mohammed, N., Biswas, G., & Goldberg, B. S. (2022). Multimodal Learning Analytics Using Hierarchical Models for Analyzing Team Performance. In *Proceedings of the 2022 Interservice/Industry Training, Simulation and Education Conference (I/ITSEC)*. National Training and Simulation Association.

Vatral, C., Mohammed, N., Biswas, G., & Goldberg, B. S. (2023a). A Framework for Performance Assessment across Multiple Training Scenarios Using Hierarchical Bayesian Competency Models. Under Review for Proceedings *of the 2023 Interservice/Industry Training, Simulation and Education Conference (I/ITSEC)*. National Training and Simulation Association (Under Review).

Villado, A., & Arthur Jr., W. (2013). The comparative effect of subjective and objective after-action reviews on team performance on a complex task. *Journal of Applied Psychology*, *98*(3), 514–528.

Wisniewski, B., Zierer, K., & Hattie, J. (2020). The power of feedback revisited: A meta-analysis of educational feedback research. *Frontiers in Psychology*, *10*, 3087.

Wright, P. C., Fields, R. E., & Harrison, M. D. (2000). Analyzing human-computer interaction as distributed cognition: The resources model. *Human–Computer Interaction*, *15*(1), 1–41.

York, T., Gibson, C., & Rankin, S. (2015). Defining and measuring academic success. *Practical Assessment, Research, and Evaluation*, *20*(1), 5.

Zhang, J., Yang, K., & Stiefelhagen, R. (2021, September). ISSAFE: Improving semantic segmentation in accidents by fusing event-based data. In *2021 IEEE/RSJ International Conference on Intelligent Robots and Systems (IROS)* (pp. 1132–1139). IEEE.

2 Instruction Intervention in Game-Based Assessment of Unmanned Systems Operator Performance

James C. Ferraro and Phillip M. Mangos

INTRODUCTION

Modern military aircraft systems are becoming increasingly autonomous, with more missions and responsibilities assigned to unmanned aerial systems (UAS; Mouloua et al., 2001). Deploying unmanned aircraft provides a number of tactical advantages, including an increase in combat efficiency and safety for aviation personnel. UAS operators encounter challenges related to workload, situation awareness, and fatigue that can vary across task domains and between aircraft platforms (e.g., MQ-9, MQ-25, and RQ-21). The advent of automated, unmanned systems has generated a host of human factors challenges, given the increased supervisory role of the human operator (see Gilson et al., 1998; Mouloua et al., 2010). Maintaining sustained attention is of primary concern, as automation has fundamentally changed certain aspects of aerial operations, particularly in the domain of UAS. Mishaps have been attributed to a failure of maintaining sustained attention during supervisory tasks of automated systems. During the period of fiscal years 1996–2006, there were 64 reported incidents with the long-haul MQ-1 Predator UAS. These include 27 Class A (greater than $1 million in damage or fatality), 3 Class B (greater than $200,000 in damage), and 34 Class C (greater than $20,000 in damage) incidents (Arrabito et al., 2010). Of the reported Class A, B, and C incidents, 62.5% were attributed to human error, specifically lack of situational awareness, as a major contributing factor. Maintaining vigilance is vital to the control of UAS, as the failure to sustain attention for an extended time period could increase the likelihood that critical signs (e.g., system malfunctions and enemy targets) are not detected in time or missed completely.

Long-haul missions performed by UAS operators ultimately pose a significant challenge for operators maintaining vigilance. Research findings support the notion that the vigilance decrement is not necessarily due to mindlessness (boredom), but rather a limitation of effortful attention (Grier et al., 2003). Vigilance-associated tasks extend beyond boredom-inducing, simple work assignments. Research on

DOI: 10.1201/9781032701639-2

behavioral and neurological measures of stress experienced during vigilance tasks strongly supports the notion that these tasks are exacting, capacity-draining assignments that impose considerable strain on cognitive resources (Warm et al., 2008). Ultimately, vigilance decrement occurs when operators must rapidly and temporarily transition from automated information processing (visual monitoring of UAS operations) to controlled information processing (e.g., response to critical system malfunction or identification of hostile activity).

Many UAS operators have transitioned from manned aviation positions to capitalize on the assumed overlap in skillsets between manned and unmanned aircraft operations. However, the gap in knowledge and skills for UAS operations from manned aviation is far greater than it may initially appear (Ferraro et al., 2017; Mouloua et al., 2019). In order to combat the performance risks, there is a need to enhance the development of attention control and other cognitive skills in UAS operators. Research strongly supports that a critical mechanism in the operator's ability to combat vigilance decrement is associated with their adaptability, in order to strategically allocate attentional resources to task components in response to novel, unpredictable, or changing task demands. Strategic attention control skills enable one to flexibly and efficiently distribute limited attentional resources in response to dynamic, unpredictable task demands, coordinate the execution of different skills, and regulate performance (Mouloua et al., 2003; Scott & Doverspike, 2005).

Simulation Training for UAS Operators

There is a push from the Department of Defense (DOD) to implement simulation-based training systems that adequately address challenges to UAS operator performance. The flexibility of simulation-based training and assessment technologies allows for an adaptable environment to rehearse mission-critical skills and identify pain points in operator performance. The ability to accurately measure—in real-time—a person's current attention control skill level provides the basis for the development of adaptive training interventions designed to customize the task's inherent attentional challenges. Additionally, a number of relatively stable individual differences have been identified as potential predictors of attention control skill development. The implication of the attention control trait concept for adaptive training is the opportunity to optimize training based on measurement of the trainee's individual trait configuration.

The DOD solicited development of "Stealth Adapt", a game-based mission rehearsal platform to measure and track user performance in several key areas of a UAS search-and-rescue (SAR) mission (Mangos, 2016). The technology was developed as a closed-loop adaptive training system, disguised as a fun and engaging game, while maintaining serious learning and adaptive training elements specifically designed to foster the development of essential cognitive skills for UAS operators.

Described within this chapter is the use of the STEALTH ADAPT training and assessment platform to test several methods of training to support user decision-making in a game-based training environment. The effort described below aimed to evaluate the efficacy of several instructional interventions aimed at increasing the training effectiveness of game-based platforms for training UAS operators. It also attempted to examine how presenting users with progressively more difficult scenarios helps to train essential skills and improve performance, as compared to random assignment of scenarios. As simulation and game-based training technologies proliferate modern military training curricula, it presents opportunities to introduce adaptable methods of support user performance and optimize their learning curve. The development and evaluation of simulated game-based training systems for unmanned systems operators can help mitigate human factors issues related to vigilance, workload, and situation awareness (Mouloua et al., 2019). Proper performance assessment and training methodology could significantly reduce instances of critical system failures, and the costs associated with them.

GAME-BASED TRAINING SOFTWARE

Adaptive Immersion's game-based STEALTH ADAPT software simulates an SAR mission, during which operators prioritize downed allies and plan a path to rescue as many as possible. The gamified SAR mission environment was built based on cognitive walkthroughs of the task domain and essential tasks with subject matter experts (SMEs), a realistic landscape that contained scattered allies in need of rescue. The virtual environments include detailed physical landscapes, first-person camera perspectives from the UAS, and models of various aircraft. This is intended to highlight the realistic representations of the environment and assets while implementing intuitive, user-tested gameplay into traditional UAS tasking.

GAME ENVIRONMENT AND GAMEPLAY

Each trial begins with a mission planning phase that tasks operators with choosing their path to each waypoint based on certain criteria: Survivability, Proximity to Weapon Engagement Zone (WEZ), and Time Until Resources are Depleted (Figure 2.1). If a waypoint's Survivability is below 15%, it should be considered a priority. Similarly, if a waypoint is within three kilometers of a WEZ (hostile territory that will damage the aircraft), it should be considered a priority. Finally, if the Time Until Resources are Depleted is below 45 seconds, that waypoint should be considered a priority. If one waypoint met all three criteria, it should be prioritized above all others, followed by a waypoint meeting two of three, and so on. This information must be considered in conjunction with the arrangement of the friendly targets and WEZs to ensure a safe and efficient execution of the task.

After confirmation of the path to rescue each waypoint, the UAS begins autonomous flight to the first waypoint. A mini-map is provided to give operators indicators of the location of each waypoint and WEZ, as well as the orientation of the aircraft relative to the direction of North. This mini-map enables operators to alter their path, selecting waypoints not intended to be next in the sequence. This may be done to avoid WEZs, or to optimize flight path to conserve resources. Additionally, the critical information about each waypoint may change over time, at which point the operator must identify how the optimal path has changed and reconsider their current trajectory. The aircraft runs on fuel that drains at a constant rate during

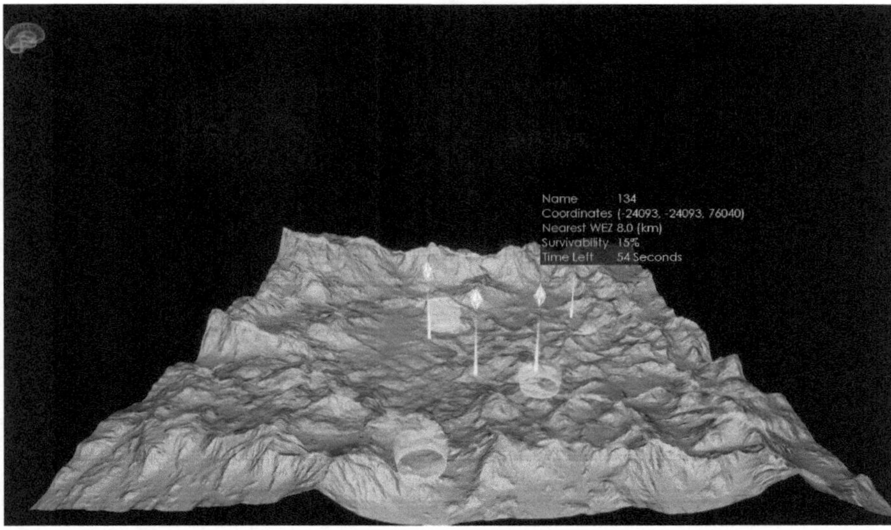

FIGURE 2.1 Mission planning gameplay interface.

flight, with a battery acting as a backup. Operators have the ability to increase the speed of the aircraft which, in turn, increases the burn-rate of the fuel and battery. Once fuel and battery are depleted, the trial is over.

Each waypoint is identified by a three-figure alphanumeric "name", or ID, to distinguish them. Authentication codes are required to collect and rescue downed allies at each waypoint. These five-figure alphanumeric codes are unique to each waypoint and are presented visually in a dialogue box presented to operators among several other, unrelated, series of messages. Operators are responsible for identifying and recalling this code to successfully pick up an ally.

The aircraft flies autonomously, meaning the operator did not have manual control over its altitude and flight path. During the session, the aircraft and operator will experience a loss of link, and the operator will be tasked with recalling certain in-flight conditions. These conditions included the in-flight view of the environment, orientation of the mini-map, name of next waypoint in the flight sequence, and authentication code for the next waypoint. A list of primary and secondary tasks is provided in Table 2.1.

Within the main display, users have access to a first-person view of the environment in front of the UAS (Figure 2.2). A mini-map is available at the top left that allows players to click on waypoints to change the path of the UAS. Enemy territory is designated as red sections on the ground, informing players of where they may take damage and letting them account for hazards during their travels. UAS resources such as health, fuel, and battery are provided adjacent to a chat window through which users receive critical information about each waypoint.

It was within this simulated environment that the training effectiveness evaluation took place.

TABLE 2.1
Search-and-Rescue Primary and Secondary Task Descriptions

SAR Tasks and Description	
Primary Task	**Description**
Rescue Downed Allies	Participants scored on the percentage of allies they are able to successfully identify and rescue
Secondary Tasks	**Description**
Waypoint Prioritization	Participants scored on their ability to properly plan their path in accordance with the rules of engagement (Mission Planning)
Loss of Link Recall	Participants scored on their ability to recall in-flight conditions after experiencing a lost connection to the aircraft
Resource Management	Participants scored on their ability to conserve fuel, battery, and health while completing their mission

EXPERIMENTAL DESIGN AND PROCEDURES

The primary objective of the training effectiveness evaluation was to objectively evaluate the potential for instructional intervention techniques within the game-based training system to enhance operator performance in the real-world mission environment. A critical requirement for training effectiveness research is to incorporate established experimental controls and procedures to provide a solid foundation

FIGURE 2.2 Mission execution gameplay interface.

for both internal and external validities of the research. One such control relates to statistical power and ensuring adequate sample size to ensure the stability of the statistical parameters of the parameters of the statistical model when the data are analyzed, ensuring appropriate inferences can be made about both reserved results, the integrity of the experimental design, and the potential for the observed patterns to translate to real-world performance improvement.

UAS operators are currently in high demand, and the availability of existing operators represents a limiting factor in the experimental design, given their extremely limited supply. A statistical power analysis was conducted to anticipate the required sample size for each of the experimental conditions. Assuming two independent variables, with two to three levels within each of the independent variables, the statistical power analysis identified a requirement for at least 15–20 observations for each parameter included in the statistical analytic model. Inclusion of additional parameters beyond the main experimental effects, such as interaction and covariate effects, increases the required sample size to achieve the criterion level of statistical power accordingly, with each additional parameter included. The inclusion of the two main effects, one interaction effect, up to five covariates (not crossed with each other or the main effects to create higher-order interaction effects) would result in the inclusion of at least eight parameters in the statistical model. Assuming no data loss, this would create a sample size requirement of at least 120–160 observations in total. In the real world, with data loss being a pervasive threat to the experimental design and statistical model integrity, one must anticipate at least 20–25% data loss for various reasons, including missing or incomplete data, careless, responding, inadequate variability in the observed performance data, and a host of other variables that could affect data quality even after extensive prescreening procedures. This placed the overall sample size requirement in the 150–200 observations range.

Subject Screening and Sampling

The scarcity of mission-ready air vehicle operators (AVOs) would render an experimental data collection employing these individuals as the primary participants extremely difficult. Our solution was to recruit a sample of individuals whose demographic characteristics, biographical history, and key experiential factors (e.g., game performance) closely resemble those of the target AVO sample. We recruited a large sample of experienced and professional gamers with substantial long-term history, of playing games of a nature and complexity similar to that of the STEALTH ADAPT system. These individuals were recruited based on their performance on a number of sequential prescreens that tested both their gaming knowledge and experience, including long-term gaming history, and regular/weekly gaming habits, frequency, and playing duration. Additionally, these individuals were matched to the prototype military unmanned pilot demographic profile for key demographic variables.

The first step in the prescreening process was a gaming knowledge test assisting familiarity with and knowledge of various gaming concepts and symbology. This step also included questions assessing long-term gaming experience history, types/genres of games played on a regular basis, and daily and weekly game performance duration, and frequency. A number of demographic variables were also captured at this initial stage of the prescreening process. In addition to basic demographic characteristics (including race, gender, age, education level, academic performance, work history, college major, and employment status, among others), information on past and present military experience was assessed for each participant. This stage of the prescreening employed a compensatory approach where either perfect performance on the knowledge portion of the assessment or near-perfect performance on the assessment combined with critical military, experiential, or other biographical history factors were used to establish a prescreening passing score for potential inclusion within the next phase of the process. The second phase of the prescreening process included screening variables for factors that could influence the quality and security of the remote gameplay experience planned for the virtual data collection. This included measured Internet upload and download speed, United States location, and geographic proximity to the servers on which the game was hosted. These first two waves of prescreens were designed to isolate the population of potential individuals eligible for participation in the study. A sample of participants was then drawn from this population and randomly assigned to the six experimental conditions. The final wave of prescreens relates only to the integrity of the data as evidence of the participants careful and conscientious response process once they were engaged in the study. This final set of prescreens includes a number of variables useful for evaluating data quality and conscientious response patterns. These include percentage of missing data on both the initial training trials and follow-up transfer trials, intra-individual variability on both the core game performance dependent variables and knowledge, self-efficacy, and perceived task difficulty surveys; univariate and multivariate outlier status on any of the primary game performance variables; illogical responses on parallel items included in the surveys assessing player attributes and motivational states; and manual review of individual response vectors for evidence of inconsistent, illogical, or careless response patterns.

A total of 2,815 participants were recruited for inclusion within the first wave of prescreens. Of these, 1,671 passed the first wave of prescreens, reflecting a pass rate of 59.4%. The next wave of prescreens resulted in an available sample of 447 (26.8%). This includes individuals who both passed the second wave of prescreens and self-selected into the study. The final wave of screening and self-selection resulted in a final sample of 179 total participants (40.0%) who completed the entire set of training trials (number of trials varies depending on the experimental condition to which they were assigned), of whom 158 completed the entire set of follow-up transfer test trials (88.7%). A summary of the total sample size by experimental conditions is provided in the table below.

A number of incentives were employed to encourage high-quality performance among the Amazon Mechanical Turk workers recruited for participation in the study. Participants were paid a base HIT payment of $25 for successful completion of the initial training HIT. Additionally, the highest performer within each daily HIT session was given a bonus of $25. Workers who completed the initial session were offered $30 to participate in the subsequent one-week follow-up HIT to complete the transfer test trials. Finally, specific instructions were included within each HIT page to ensure conscientious adherence to all instructions and consistently high effort throughout the entire set of trials. This included knowledge checks within the initial instructions that required participants to accurately different elements of the display to demonstrate the familiarity of being able to move ahead with the actual training trials.

Instruction Interventions and Design

To begin the study, participants were randomly assigned to one of two Instructional Strategy conditions. In one of these conditions, participants receive pre-mission guidance in the form of an instructional intervention, while in the other they did not. This intervention provided participants with additional hints to help with mission planning. Participants were also assigned to one of three training intervention conditions. In one condition, training scenarios were presented to participants in a randomized order. Each scenario, scored with a difficulty rating determined by mission conditions, was uniquely rated 1 through 10. In another training condition, scenarios were presented to get progressively more difficult as the participant advances. This sequence was designed specifically to mimic the adaptive training algorithm developed for STEALTH ADAPT. In a final training condition, the participant did not receive any training scenarios. The six conditions are outlined in Table 2.2.

Participants began a session with a series of surveys. The first survey gathered general demographic information (age, gender, education level, etc.) as well as information regarding participants' experience playing videogames. A self-efficacy scale was provided to assess participants' attitudes toward their capabilities in the tasks required in the STEALTH ADAPT missions (e.g., task prioritization). Finally, the Dundee Stress State Questionnaire (DSSQ) was administered to assess pre-task engagement, distress, and worry.

Participants in the Randomized and Progressive Training conditions experienced ten training trials following the pre-trial survey. These trials were presented in an order determined by their Training condition. Depending on their Instruction Strategy condition, they either received the instructional intervention or not. They then completed two experimental test trials. These final trials were the same for all participants, regardless of condition. Participants in the No Training conditions advanced from the pre-task surveys directly to the two experimental test trials. Following these two trials, participants completed a post-task DSSQ.

There was a one-week hiatus between sessions, and participants were brought back to complete additional experimental trials. During this second session, participants

TABLE 2.2
Conditions of the Training Effectiveness Evaluation Study

Group 1	**Group 2**	**Group 3**
Randomized Training/ No Instruction Intervention	Progressive Training/ No Instruction Intervention	No Training/ No Instruction Intervention
Group 4	**Group 5**	**Group 6**
Randomized Training/ Instruction Intervention	Progressive Training/ Instruction Intervention	No Training/ Instruction Intervention

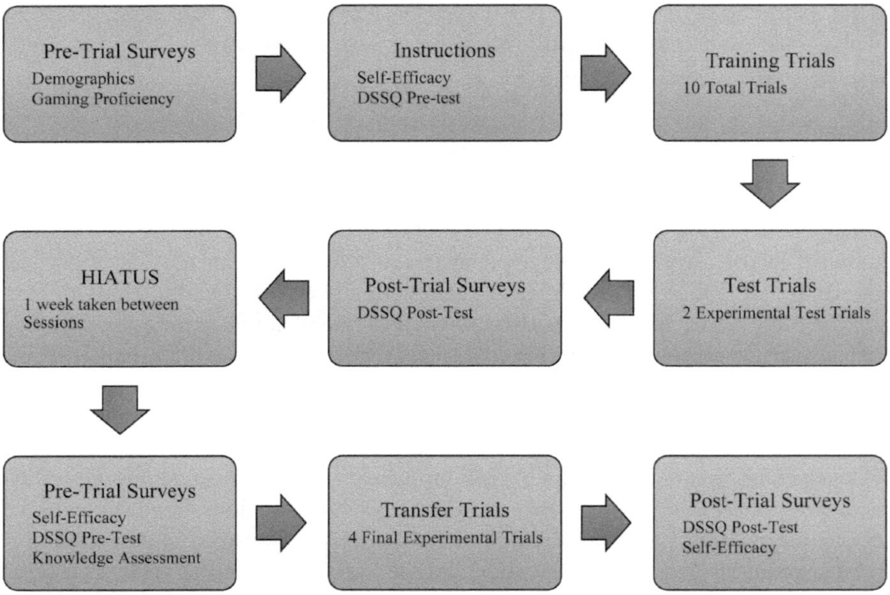

FIGURE 2.3 Research study design sequence.

once again began by completing the DSSQ. Additionally, prior to the additional trials intended to assess transfer of training, participants completed a knowledge assessment to establish what information regarding the interface and rules of engagement they retained from their training session.

All participants then completed the same four final transfer trials. These transfer trials were designed to mimic real-world operational conditions, with a significant increase in difficulty, and incorporating unanticipated events (e.g., Loss of Link and authentication code decryption requirement) that were never experienced during training or the instructional sessions. This is consistent with the notion of "truly surprising events", where the ultimate indicator of transfer of training is effective performance under real-world conditions never experienced during training. Two of the transfer trials were the same as the final two experimental test trials as the previous session, and two were unique and never seen previously by participants. These trials were then followed by the DSSQ post-test and a final assessment of participants' self-efficacy. This procedure is shown in Figure 2.3.

ASSESSMENT OF TRAINING INTERVENTION IMPACTS

The training effectiveness study was conducted over the course of six weeks in November and December 2021. The 179 participants who progressed through three stages of prescreens completed their initial training trials during the first session, and 158 of these completed the second set of transfer test trials during the second

TABLE 2.3
Participant Pool Breakdown

Condition	Sample Size—Training	Sample Size—Transfer
Instruction	94	81
Random Training	36	29
Progressive Training	32	27
No Training	26	25
No Instruction	85	77
Random Training	26	23
Progressive Training	29	27
No Training	30	27
Total	179	158

session. A breakdown of the number of participants in each experimental group is provided in Table 2.3.

Key demographic and cognitive state variables were assessed, including gender (65.54% Male, 34.46% Female), age (M = 35.70, SD = 7.94), previous night's sleep hours (M = 7.08, SD = 1.18), color vision (100% normal color vision), education level (100% high school level or above, 51.96% Bachelor's degree or above), prior military and flight experience (8.00%), and academic GPA (76.54% above 3.0).

Performance in both the training and transfer conditions was assessed using mathematical modeling techniques to produce a quantitative score for each performance dimension, at the level of each individual training or transfer trial, and then normalized across the entire sample within the individual trial to produce a standardized score for each variable with a mean of 0 and standard deviation of 1.

It was expected that the Progressive Training intervention would prove beneficial and result in higher scores across performance measures in the transfer trials. It was also expected to result in better performance in knowledge assessment items and higher self-efficacy. The presence of the instructional intervention (Instruction condition) was also expected to improve performance, knowledge, and self-efficacy.

Statistical Analysis and Results

Among the multitude of individual performance variables available for use, we analyzed and reduced a set of candidate metrics determined to be representative of the players' accuracy, efficiency, and effectiveness, and that were relatively oblique metrics (i.e., having low correlations with each other), to provide a comprehensive and nonredundant picture of the players' overall performance both during training and transfer.

The shortlisted performance variables considered for further analysis are included in Table 2.4.

A series of growth curve analyses were run on the training scenario data to examine participants' performance trends over time. The area beneath the growth curve

TABLE 2.4
Measured Performance Variables and Corresponding Phase of Measurement

Performance Metric	Phase of Measurement
Situation Awareness Circumplex	Training
Situation Awareness Circumplex	Transfer
Average Speed Improvement	Training
Average Idle Time Improvement	Training
Average Attempts per Waypoint	Training
Average Prioritization Score	Training
Average Non-Idle Speed	Training
Average Prioritization Score	Transfer
Average Idle Time	Transfer
Average Attempts per Waypoint	Transfer
Knowledge Assessment	Transfer
Self-Efficacy	Training (1)
Self-Efficacy	Training (2)
Self-Efficacy	Transfer

was estimated and then normalized to examine prior performance compared to the transfer trials to assess the cumulative level of improvement. The purpose of this analysis was to measure the error-corrected cumulative performance growth during training, reflecting level and rate of change in initial skill acquisition. These performance growth trajectories can then be used to predict both pre-retention (immediate) transfer performance and post-retention transfer performance (i.e., after the one-week retention interval).

A series of ANCOVA and regression analyses were performed to examine main effects or interaction effects present between conditions, particularly to the extent to which an IV may have affected performance between the initial test trials and the transfer trials. Demographic information and gaming experience were utilized as covariates.

Core Performance Variable Selection and Dimension Reduction

The STEALTH ADAPT training system was designed within a performance-based measurement architecture, incorporating a wide variety of fine-grained performance metrics and mathematical models to capture the full spectrum of the player's performance. These individual metrics can be observed within single time instances, ranging from within-session play time intervals (measured in seconds) to performance within individual training or transfer trials, to individual growth patterns observed across the entire training study. The building blocks of this performance measurement strategy include individual metrics reflecting the players observable performance vis-à-vis the game rules, scenario objectives, and rules of engagement. These reflect how accurately and efficiently the player is operating the simulated UAS, the efficiency with which they are burning consumable resources such as fuel and battery, the speed and efficiency of navigating the mission environment, prioritization accuracy related to the pattern of rescuing the downed friendly forces, memory and data management for accomplishing critical mission objectives, situational awareness related to where the player is in their intended movement and actions at a given point in the mission, their knowledge of the rules of engagement, and their experienced stress and confidence levels in accomplishing the mission objectives.

We evaluated the bivariate correlations among these candidate variables, and performed additional dimension reduction in the form of factor analyses to identify the most promising and relatively uncorrelated variables for inclusion as the court-dependent variables in the training effectiveness study analysis (Table 2.5).

Factor analysis was then performed, using principal axis factoring, with direct oblimin rotation of the factor solution (an oblique rotation technique allowing factor solutions to be correlated). This was designed to derive the reduced set of latent variables underlying the 14 candidate performance indicators, associate indicators with latent variables, and use these results to identify the most promising dependent variables for the core effectiveness evaluation analysis. The factor analysis derived three latent factors with eigenvalues greater than 1.00. The absolute values of the factor loadings ranged from 0.01 to 0.58, with on average 3 to 4 indicators loading on to each of the derived factors.

TABLE 2.5
Candidate Variable Correlation Values

	Performance Metric	1	2	3	4	5	6	7	8	9	10	11	12	13	14
1	Situation Awareness Circumplex - Training	--													
2	Situation Awareness Circumplex - Transfer	.316**	--												
3	Average Speed Improvement - Training	-0.007	0.087	--											
4	Average Idle Time Improvement - Training	-0.055	-0.128	.302**	--										
5	Average Attempts per Waypoint - Training	-0.051	-0.065	.192*	.476**	--									
6	Average Prioritization Score - Training	-0.151	0.084	-0.019	-0.144	0.146	--								
7	Average Non-Idle Speed - Training	-0.066	-0.044	.249*	0.017	-0.139	-0.132	--							
8	Average Prioritization Score - Transfer	-0.042	-0.025	0.058	-0.003	0.003	0.093	-0.050	--						
9	Average Idle Time - Transfer	-.336**	-.453**	-0.044	.206*	-0.029	0.014	-.291**	0.044	--					
10	Average Attempts per Waypoint - Transfer	0.067	0.117	-0.136	-0.137	-.217*	-0.055	-0.075	-0.054	-0.067	--				
11	Knowledge Assessment - Transfer	.195**	.201*	0.068	-0.086	0.114	-.238**	-0.028	-0.107	-.301**	0.039	--			
12	Self-Efficacy - Training 1	0.013	-0.045	-0.063	-0.127	-0.077	0.049	0.016	-0.022	0.095	-0.018	-0.092	--		
13	Self-Efficacy - Training 2	.148*	0.071	0.069	-0.147	-0.046	-0.074	0.021	0.023	0.069	0.032	.568**	.195**	--	
14	Self-Efficacy - Transfer	0.131	0.066	0.106	-0.020	-0.046	-0.054	0.053	.173*	-0.128	0.066	.481**	0.146	.791**	--

Growth Curve Prediction

One critical aspect of training performance of interest in this experiment was the overall degree and pattern of performance improvement within the training trials, that is, the participants' individual learning curves. The overall level of performance improvement refers to the Delta between the standardized performance scores of the first and final trials. The pattern of performance change corresponds to whether the level of change is consistent from trial to trial, forming a linear growth trend, or if there is significant acceleration or curvature at key points in the training process, which could be assessed by fitting a polynomial, logarithmic, or exponential growth curve to the intra-individual performance data.

We estimated both polynomial (quadratic) and exponential growth functions for each individual's normalized training performance data for each core performance variable. This reflects their initial level of performance (intercept), as well as the level and rate of growth over time, either positive or negative, as expressed in the quadratic or exponential function. This model allowed us to account for the time related to the number of trials which varied depending on training condition.

$$z\left(log\left(performance\,score\right)\right)=slope*exp\left(trials\right)+intercept.$$

In this model, the intercept represents initial performance at the beginning of the training trials, while the slope corresponds to the rate of change of performance improvement across the trials. The full exponential model was calculated for each of the participants across conditions except for the "no training" condition which immersed participants immediately in the training test trials. An intercept-only form of the model was calculated for those participants with inadequate growth curve data. To construct the growth curve metrics, we needed at least three data points across the ten trials of data to construct a reliable trajectory of performance. Therefore, individuals with fewer than three training trial data points were eliminated from the data set.

After estimating the best fitting exponential function for each person's data, we then estimated the area under the growth curve. The area under the exponential curve reflects the individual's cumulative performance improvement, assuming the potential for quadratic or exponential growth change. The area under the curve was then standardized into a *z*-score. This final *z*-score captures the totality of each individual's objective training performance over the training trials. The final objective metric has a mean of 0.00 and a standard deviation of 1.00.

This approach was used to model performance improvement over the training trials. The level and pattern of improvement in and of itself represent a meaningful dependent variable. Additionally, the reliable variance of the performance scores measured in the transfer trials represents a critical indicator of overall learning. Together, these metrics can be used to form a comprehensive statistical model incorporating the core manipulated independent variables and covariates as predictors of cumulative learning and resulting transfer performance.

Every performance measure was standardized using a single *z*-score calculation, normalizing the measure across all individuals (across external conditions) within an individual learning trial. This allowed us to make meaningful, apples-to-apples comparisons across trials when more difficult scenario conditions raised or lowered the effective performance ceilings across trials. This also allowed us to construct meaningful learning curves across trials based on intra-individual performance change.

Average Attempts per Waypoint Efficiency

The primary dependent variable related to the player's ability to rescue the downed friendly forces was the average number of attempts per waypoint. This was used as a proxy for the actual number of waypoints, which differed by trial, and was subject to ceiling effects as the time limitation imposed by available fuel allowed most participants to reach 100% rescue for most trials before running out of fuel (i.e., with some spare fuel to return and reattempt any missed waypoints). The average number of attempts gets at the player's efficiency and prioritization effectiveness, and provides an invariant scale to compare performance across trials.

Results indicated a significant time (trial) effect for an average number of attempts per waypoint across the ten training trials, reflecting fewer attempts per waypoint (12%) and greater efficiency with practice (94% in Trial 1, 82% in Trial 10).

We observed a significant effect for training condition on the normalized number of attempts per waypoint training performance growth metrics (higher scores reflecting better improvement in player efficiency). Individuals in the Progressive Training condition demonstrated better growth in their efficiency relative to individuals in the Random Training condition by approximately 0.50 SD units (Progressive Training $M = 0.28$, $SD = 0.96$; Random Training $M = -0.27$, $SD = 0.96$; $F(1, 119) = 9.77$, $p < .01$) (Figures 2.4 and 2.5).

Average Idle Time Efficiency Metric

The average idle time efficiency metric is the average amount of time spent in idle positioning over waypoints as players validate and enter authentication codes, and is a measure of memory performance and data management efficiency. Lower idle times translate into faster and more efficient rescues. This metric was normalized for meaningful cross-trial comparisons. The normalized training improvement performance metric reflects the level of improved efficiency (i.e., reduced idle time) across the training trials, so higher scores mean better overall improvement over time. The raw version of this metric is the actual average amount of idle time, whereas the transformed efficiency version of this metric is reversed such that higher scores reflect better performance.

Growth curve modeling of the raw idle time score showed a significant, negative quadratic growth pattern, with high average idle times in the first two trials quickly reducing across the subsequent trials. Additionally, within-trial variability decreased progressively across the trials as participants' rescue strategies and idle times stabilized over time (Figure 2.7).

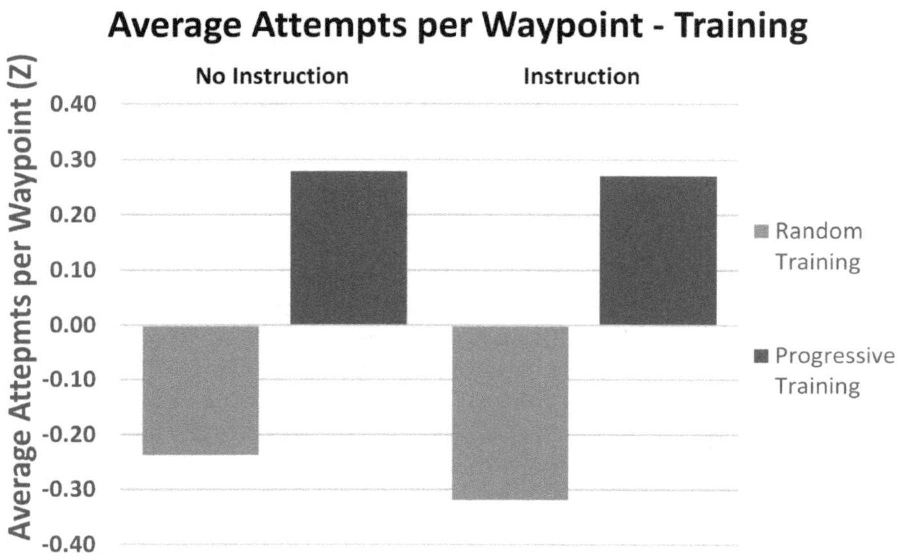

FIGURE 2.4 Average attempts at rescue per waypoint by instruction and training group.

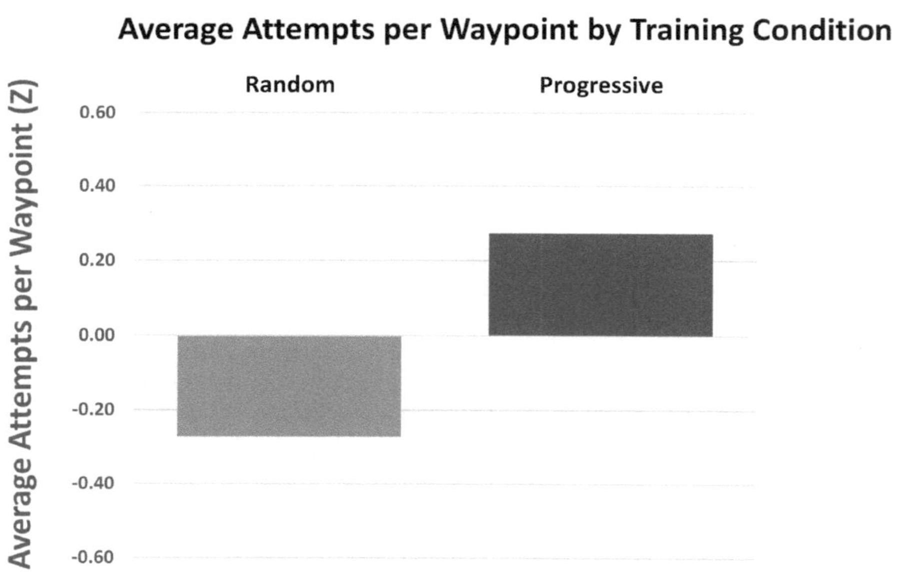

FIGURE 2.5 Average attempts at rescue per waypoint by training group.

Results for the average attempts efficiency metric during the follow-up transfer trials showed no significant effect for either Training ($F(2, 152) = 1.00$, ns) or Instruction condition ($F(1, 152) = 90.39$, ns) (Figure 2.6).

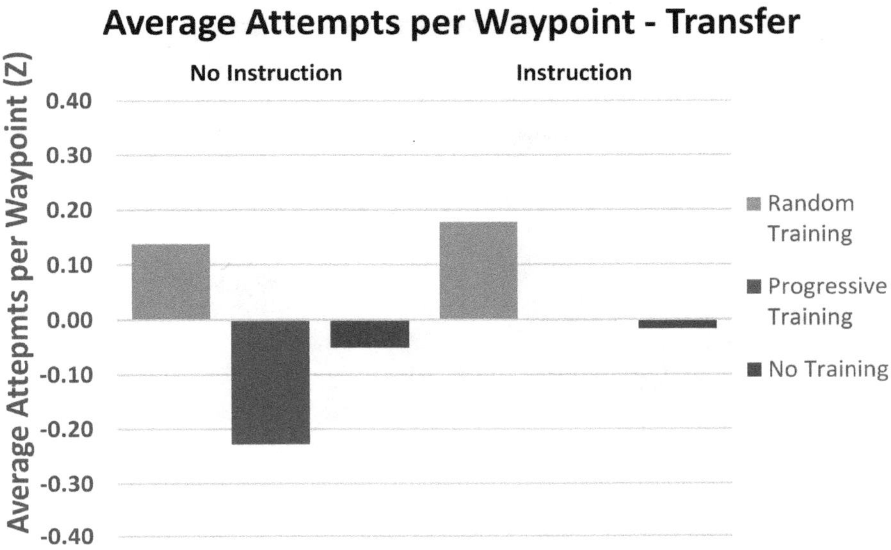

FIGURE 2.6 Average attempts per waypoint in the follow-up transfer trials.

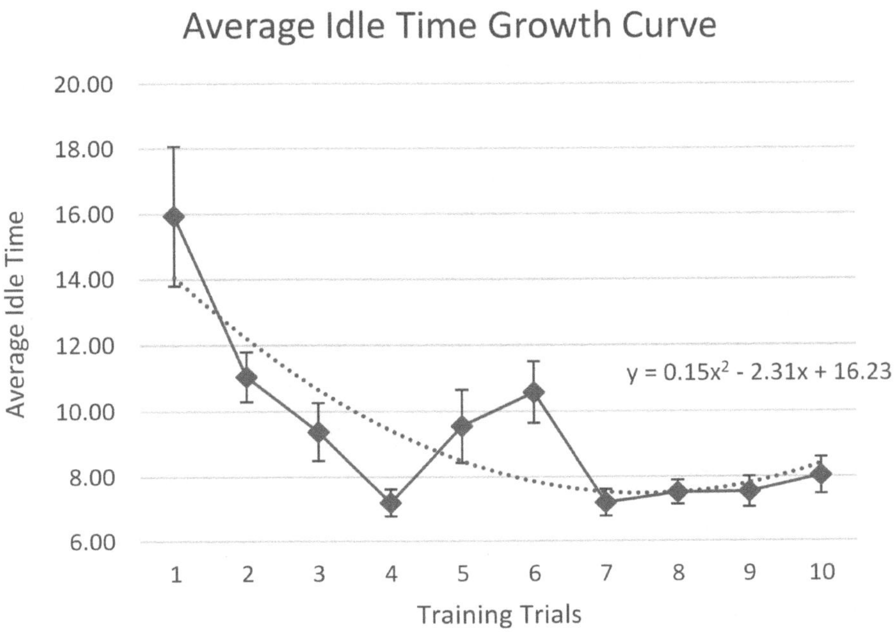

FIGURE 2.7 Growth curve showing the average idle time across training trials.

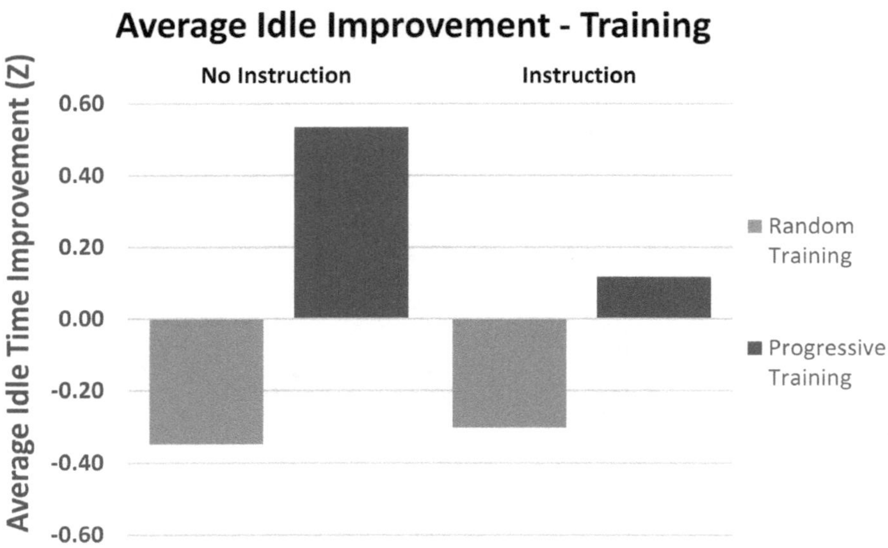

FIGURE 2.8 Average idle time improvement in training trials.

Results with respect to the efficiency metric showed a significant effect for training condition on the normalized idle time training performance growth metric (higher scores reflecting better improvement in player efficiency). Individuals in the Progressive Training condition demonstrated better growth in their efficiency relative to individuals in the Random Training condition by approximately 0.70 SD units (Progressive Training $M = 0.34$, $SD = 1.18$; Random Training $M = -0.33$, $SD = 0.64$; $F(1, 119) = 14.39$, $p < .01$) (Figures 2.8 and 2.9).

The improvement in efficiency observed across training translated into better performance during the transfer trials, with individuals in the Progressive Training condition demonstrating the lowest average idle time relative to the other conditions by a factor of 0.40 SD units (Random Training condition) and 0.30 SD units (No Training condition). This pattern approached but did not achieve statistical significance ($F(2, 152) = 2.11$, $p = .13$) (Figure 2.10).

Average Speed Performance Metric

The average speed metric is the average speed of the player's non-idle time travel during mission execution, that is, the relative speed at which they are rescuing the friendly targets. This factors out the idle time required to recall authentication codes and is a measure of overall mission efficiency. This metric was normalized to produce meaningful comparisons across trials and establish growth curves based on an identical metric.

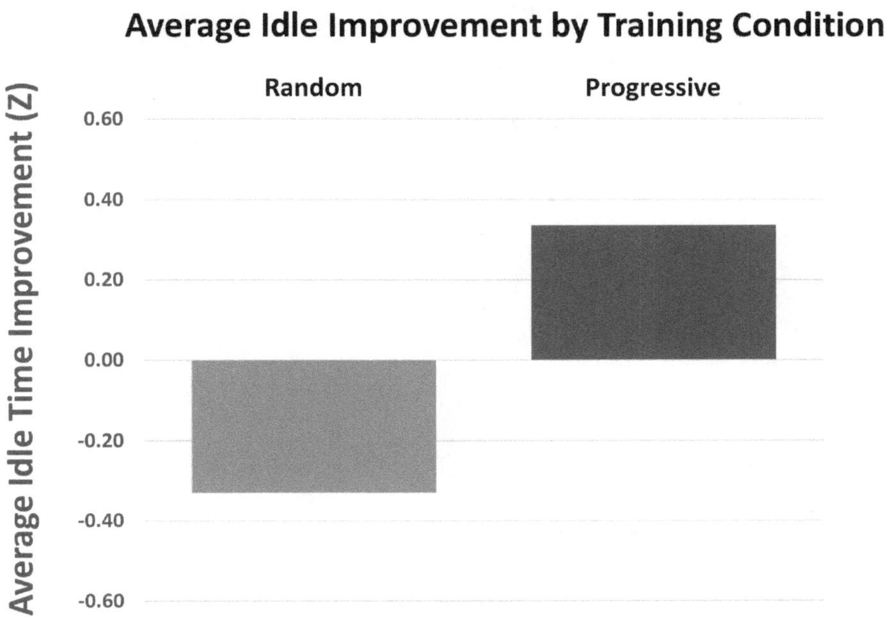

FIGURE 2.9 Average idle time improvement between training groups.

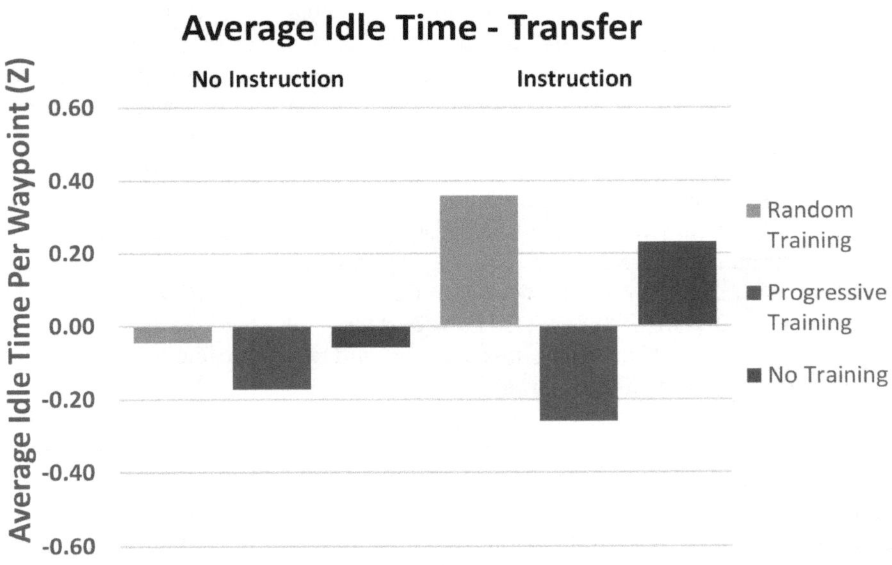

FIGURE 2.10 Average idle time in the transfer trials.

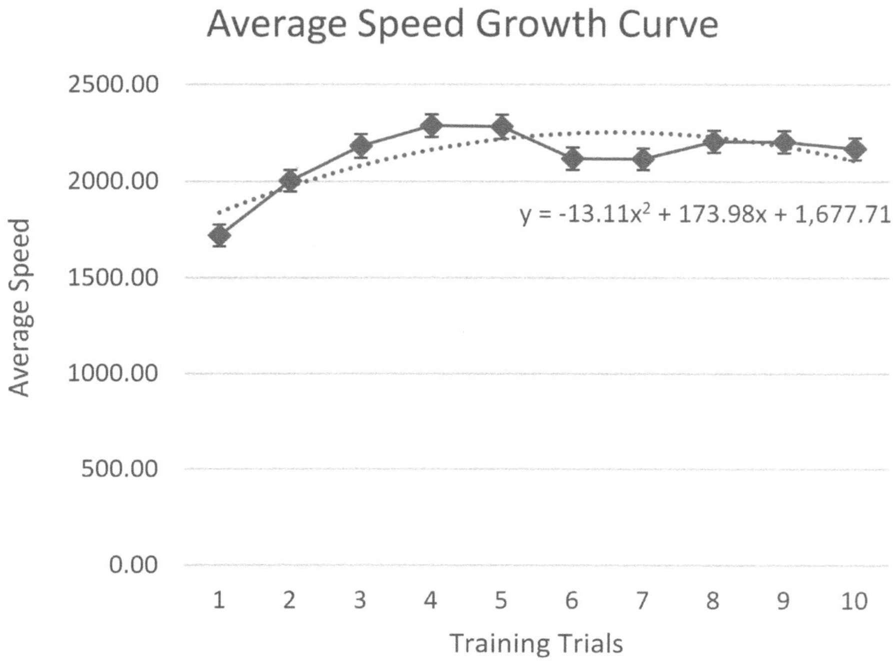

FIGURE 2.11 Growth curve showing the average UAS speed across training trials.

Results indicated substantial improvement in the players' average speed performance during training trials, and that this improvement was contingent on training condition. There was a significant, positive linear trend, and negative quadratic trend showing accelerated growth across the first half of the training trials which then leveled off across the second half. This trend shows that players were better able to manage the UAS speed to their advantage and more consistently operate at higher speeds across the later trials (Figure 2.11).

The Progressive Training condition resulted in greater improvement gains in speed performance relative to the Random Training condition by a factor of nearly a full standard deviation unit (Progressive Training $M = 0.53$, $SD = 1.12$; Random Training $M = -0.43$, $SD = 0.84$; $F(1, 119) = 24.29$, $p < .01$) (Figure 2.12).

However, this performance improvement did not translate into enhanced speed performance during the transfer sessions ($F(2, 152) = 0.71$, ns). This may be related to the inherently different speed requirement during the transfer trials which included considerably higher quantities of waypoints and a need to maintain higher levels of speed throughout. This, in turn, reduced the variability in the participants' speed performance metric during the training trials, thereby limiting the potential effects of the experimental variables on this metric (Figure 2.13).

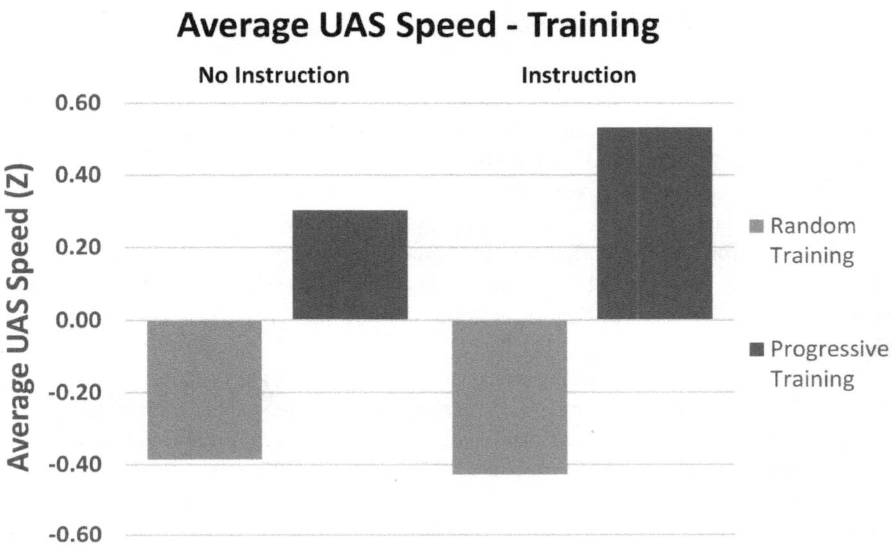

FIGURE 2.12 Average UAS speed between instruction and training conditions during training trials.

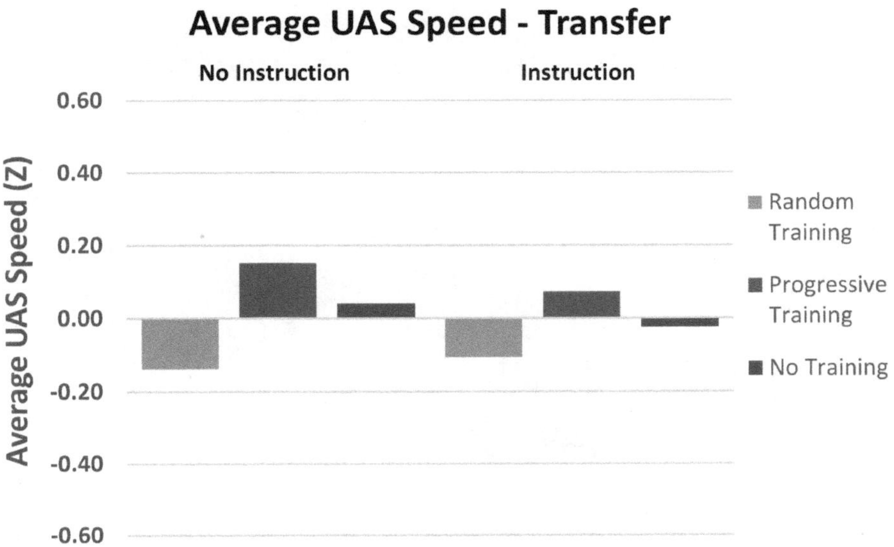

FIGURE 2.13 Average UAS speed between conditions during transfer trials.

Target Prioritization Performance

The players' target prioritization performance was measured as the Euclidean distance value between the player's sequence of waypoint rescues and the optimal sequence based on the rules of engagement and ground truth priority values. This value was then normalized within each trial, providing a critical means of comparison across trials, given that the raw Euclidean distance values naturally increase with larger numbers of waypoints, and therefore, greater numbers of opportunities for suboptimal prioritizations. This metric relates directly to the player's mission planning abilities and adherence to the specific rules of engagement, as the waypoint sequencing is done primarily during the mission planning phase.

Results indicated substantial variability in the players' prioritization performance during the training trials, as well as in their improvement on this metric across the trials. This level of improvement was contingent on the experimental variables, with a significant main effect of Instruction condition (Instruction $M = 0.23$, $SD = 0.99$; No Instruction $M = -0.18$, $SD = 0.97$; $F(1, 119) = 5.18$, $p < .05$), marginal main effect for training condition (Progressive Training $M = 0.12$, $SD = 1.05$; Random Training $M = -0.11$, $SD = 0.94$; $F(1, 119) = 2.11$, $p = .15$), and a significant instruction × training condition interaction effect ($F(1, 119) = 6.05$, $p < .05$). The form of this interaction indicates that though the Progressive Training condition generally produced greater prioritization performance improvement across the trials, relative to the Random Training condition, the training condition benefit was substantially greater when instruction was also provided. In this sense, the instructional intervention essentially boosted the beneficial effect of Progressive Training (Figure 2.14).

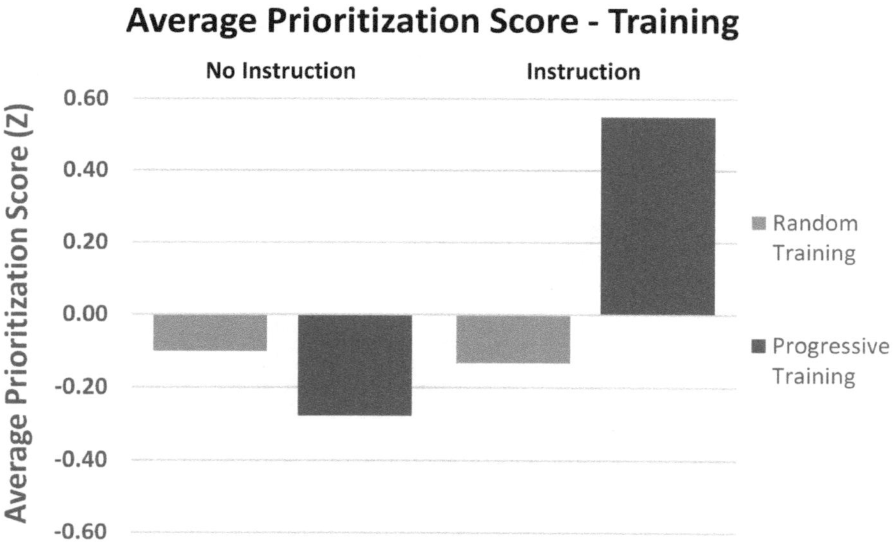

FIGURE 2.14 Average target prioritization score between conditions in training trials.

The training condition benefit enhanced performance during the transfer trials as well. There was a significant main effect for training condition, resulting in a 0.50 standard deviation task prioritization performance enhancement in the Progressive Training versus Random Training Condition (Progressive Training $M = 0.25$, $SD = 0.91$; Random Training $M = -0.27$, $SD = 1.01$; $F(1, 152) = 3.41$, $p < .05$) (Figures 2.15 and 2.16).

Knowledge Assessment

The knowledge assessment was presented during the follow-on transfer trial sessions, designed to assess understanding and retention of the key gameplay concepts and rules of engagement. Scores on the knowledge assessment provide a complementary perspective on the participants' transfer performance, providing a more robust picture as to whether performance improvements are due to a deeper understanding of the gameplay concept as opposed to merely experience-driven improvements in speed and efficiency.

Results with respect to knowledge assessment revealed that although neither independent variable's main effects were significant, there was a significant instruction × training condition interaction ($F(2, 173) = 3.61$, $p < .05$). The form of this interaction indicates that the gap in the mean knowledge assessment scores between the Instruction and No Instruction conditions widened in the Random Training and No Training conditions, and was reduced in the Progressive Training condition. This provides evidence for a synergistic effect when the instructional intervention is provided in conjunction with the Progressive Training, which is designed

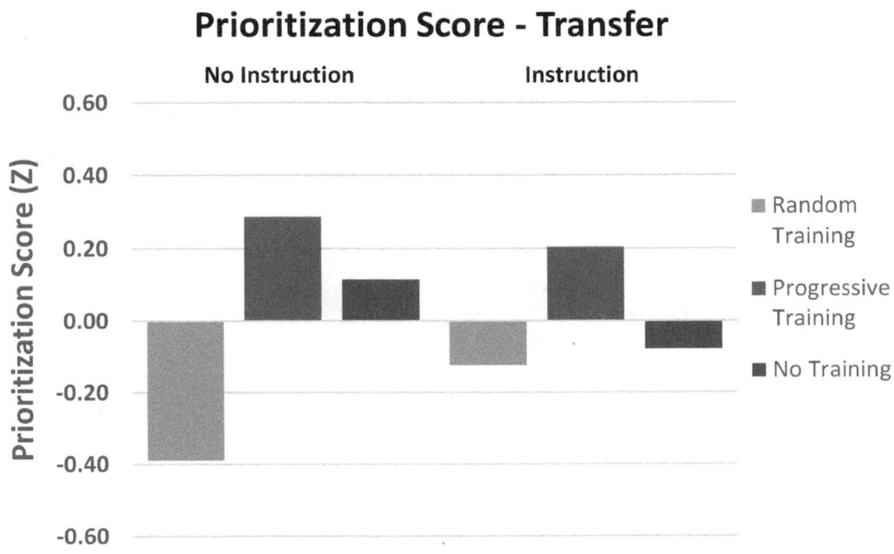

FIGURE 2.15 Average target prioritization score between conditions in transfer trials.

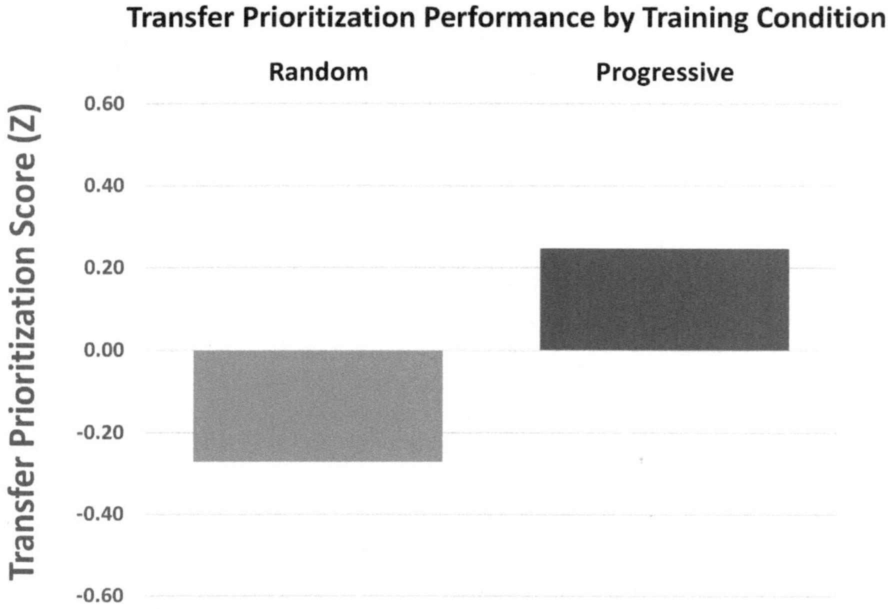

FIGURE 2.16 Target prioritization scores between training groups.

to mimic AIT's adaptive training algorithm. Instruction actually had a deleterious effect on performance in the No Training condition, potentially causing confusion in the absence of opportunities to practice the instructional concepts provided during the designated training trials. In the Random Training condition, the instructional intervention provided an approximately 0.57 standard deviation unit increase in performance ($M = -0.31$ versus 0.26 in the No Instruction versus Instruction conditions, respectively), whereas in the Progressive Training condition, the instructional intervention produced a smaller (0.10 standard deviation unit) increase ($M = -0.05$ versus 0.05 in the No Instruction versus Instruction conditions, respectively). This speaks to the compensatory nature of the training intervention, specifically for adaptive training to serve as a proxy for explicit instructional content. The Progressive Training intervention raised the overall group mean across the instructional conditions by a marginal amount (0.10 standard deviation units), compared to the Random Training condition, although the No Training condition produced a marginally higher group mean (with the Instruction versus No Instruction mean values in the reverse direction) (Figure 2.17).

Self-Efficacy

Results with respect to the self-efficacy measure provided additional evidence of the joint benefit of the training and instruction interventions. Reliable

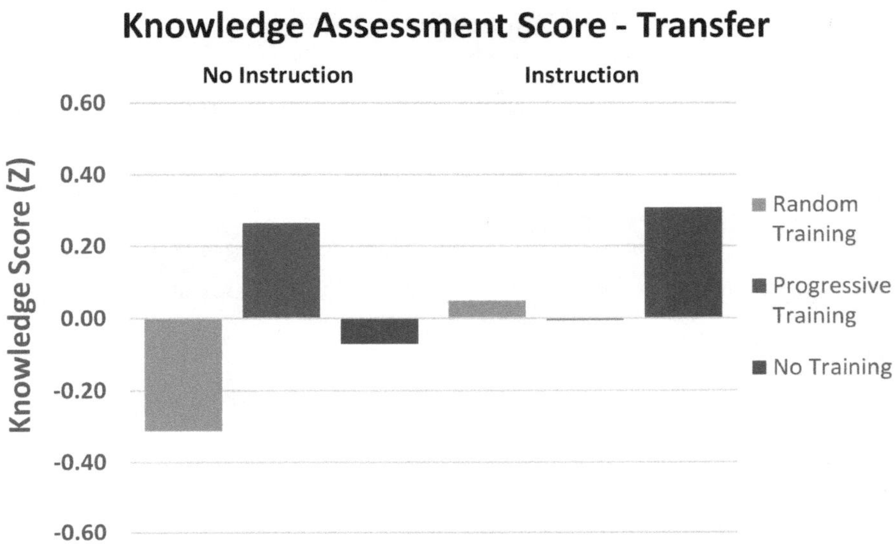

FIGURE 2.17 Knowledge transfer scores between conditions in transfer trials.

experimental effects were observed for both the training ($F(2, 173) = 2.40, p < .10$) and instruction variables ($F(1, 173) = 3.73, p < .10$). These values approached but did not exceed the $p < .05$ statistical significance criterion, but would achieve significance with a slight increase in sample size and statistical power. However, the pattern of results reflected beneficial effects of both training intervention with a consistent group mean advantage for the Instruction versus No Instruction condition (approximately 0.30 standard deviation units across the three training conditions), as well as consistently large spread across the three training conditions, with an approximately 0.50 standard deviation improvement offered by Progressive versus Random Training, but slightly better performance observed in the No Training condition (Figure 2.18).

The training effectiveness evaluation provides a critical component to the validation and future development of the training platform. The results of this effort will inform areas of need for future iterations of the software, and additional meetings with SMEs will help create additional training scenarios for future platform validation studies. The results showed that the intended interventions (Progressive Training/Instruction), designed to mimic the adaptive training algorithm of the STEALTH ADAPT system, individually or synergistically produced tangible performance benefits in a mission-realistic transfer environment. This trend was reliably reproduced across several individual performance metrics reflecting player accuracy, efficiency, and mission effectiveness. Additionally, these results extended to several non-performance variables reflecting player motivational, affective, and knowledge states.

FIGURE 2.18 Self-efficacy scores between conditions in transfer trials.

INSIGHTS AND PRACTICAL IMPLICATIONS

The findings from this training effectiveness evaluation offer several important insights for the development and implementation of game-based training systems, particularly for complex operational environments like unmanned systems control. The results demonstrate that progressive, adaptive training approaches—when combined with targeted instructional interventions—can produce meaningful improvements in operator performance across multiple dimensions, including task efficiency, prioritization accuracy, and knowledge retention.

Several key implications emerge from these findings. First, the superior performance of participants in the Progressive Training condition suggests that carefully structured, scaffolded difficulty progression is more effective than randomized exposure for developing core operational competencies. This aligns with established learning theory regarding the importance of scaffolded skill development (Warm et al., 2008). The effectiveness of Progressive Training observed here parallels findings from other high-stakes operational domains such as air traffic control (ATC), where research has shown that graduated exposure to increasing traffic complexity leads to better controller performance compared to random presentation of scenarios.

The synergistic effect between Progressive Training and instructional intervention is particularly noteworthy. When explicit instruction was combined with Progressive Training, participants showed enhanced prioritization performance and knowledge retention compared to either intervention alone. This suggests that optimal training outcomes may require both structured skill-building opportunities and clear conceptual guidance—a finding that has important implications for training system design across multiple domains, including manned aviation, process control, and military command and control operations.

The transfer of training effects to novel, more challenging scenarios is especially promising from an applied perspective. The fact that performance improvements persisted even in substantially more difficult transfer trials suggests that the training interventions fostered genuine, flexible, transferable skill development rather than just rote familiarity with specific scenarios. This kind of robust transfer is crucial for operational domains where personnel must be prepared to handle unexpected situations and novel challenges.

Several directions for future research and development emerge from these findings:

1. Investigation of individual differences in training responsiveness, particularly examining how factors like prior gaming experience, spatial ability, and working memory capacity may moderate the effectiveness of different training approaches.
2. Development of more sophisticated adaptive algorithms that can dynamically adjust both scenario difficulty and instructional support based on real-time performance metrics.
3. Exploration of additional performance dimensions such as team coordination and communication, which are increasingly important in modern unmanned systems operations.
4. Integration of physiological monitoring to better understand operator workload and attention states during training, potentially enabling more precise adaptation of training difficulty.
5. Extension of the training approach to other operational domains such as cyber operations, intelligence analysis, and emergency response, where similar cognitive demands for sustained attention and complex decision-making exist.

The parallel between UAS operation and other complex operational domains suggests broader applications for this training approach. The challenge of maintaining vigilance while monitoring largely automated systems is increasingly common across numerous military and industrial domains. In military aviation, both manned and unmanned platforms are experiencing increasing levels of automation, requiring pilots and operators to transition effectively between automated and manual control modes. The findings regarding Progressive Training and sustained attention have direct implications for training programs across the spectrum of aviation platforms, from traditional fighter aircraft to maritime patrol aircraft to emerging autonomous combat aircraft.

ATC represents another domain where the findings have clear applications. Controllers face similar challenges in maintaining situation awareness while monitoring multiple automated systems and coordinating multiple aircraft. The demonstrated effectiveness of progressive, game-based training in developing sustained attention and decision-making skills aligns well with emerging needs in ATC training, particularly as the National Airspace System becomes more automated and controllers must manage an increasing mix of manned and unmanned traffic.

Beyond aviation, the training approach shows promise for other unmanned systems domains, including ground robotics, maritime systems, and space operations. Each of these domains requires operators to maintain vigilance while supervising automated systems, often for extended periods. The success of Progressive Training in developing sustained attention and complex decision-making skills could inform training programs for autonomous ground vehicle operators, unmanned surface and subsurface vessel controllers, and satellite operations personnel.

In the industrial sector, similar cognitive demands exist in process control operations, power plant management, and automated manufacturing supervision. These domains share key characteristics with UAS operations, including the need to maintain situation awareness during largely automated operations while being prepared to respond quickly to anomalies or emergencies. The study's findings regarding the effectiveness of combined Progressive Training and instructional intervention could help inform the training program design for these industrial applications, particularly as facilities become more automated and operator roles shift increasingly toward system supervision.

Moreover, the finding that game-based training can effectively develop complex operational skills has implications for recruitment and selection. The successful use of experienced gamers as proxy participants suggests potential value in considering gaming proficiency as one indicator of aptitude for certain operational roles, though this would require careful validation.

As unmanned systems continue to proliferate across military and civilian applications, the need for effective operator training will only increase. The insights gained from this study suggest that game-based training platforms, when properly designed with progressive difficulty and integrated instruction, can play a valuable role in meeting this growing training demand. Future development should focus on further refinement of adaptive algorithms, expansion of scenario complexity, and integration with other training modalities to create comprehensive preparation for operational challenges.

REFERENCES

Arrabito, G. R., Ho, G., Lambert, G., Rutley, M., Keillor, J., Chiu, A.…Hou, M. (2010). Human Factors Issues for Controlling Uninhabited Aerial Vehicles: Preliminary Findings in Support of the Canadian Forces Joint Unmanned Aerial Vehicle Surveillance Target Acquisition System Project (Technical Report No. DRDC Toronto TR 2009-043). Toronto, Canada: DRDC Toronto.

Ferraro, J., Clark, L., Christy, N., Mouloua, S. A., Mangos, P., & Mouloua, M. (2017). Effects of adaptive training on search performance and workload of unmanned autonomous systems. Proceedings of the 61st Annual Meeting of the Human Factors and Ergonomics Society.

Gilson, R., Richardson, C., & Mouloua, M. (1998). Key human factors issues for UAV/UCAV mission success. AUVSI'98.

Grier, R. A., Warm, J. S., Dember, W. N., Matthews, G., Galinsky, T. L., Szalma, J. L., & Parasuraman, R. (2003). The vigilance decrement reflects limitations in effortful attention, not mindlessness. *Human Factors: The Journal of the Human Factors and Ergonomics Society*, *45*, 349–359.

Mangos, P. (2016). *Stealth Adapt [Computer Software].* Tampa, FL: Adaptive Immersion Technologies.

Mouloua, M., Gilson, R., Daskarolis-Kring, E., Kring, J., & Hancock, P. (2001, October). Ergonomics of UAV/UCAV mission success: Considerations for data link, control, and display issues. In Proceedings of the Human Factors and Ergonomics Society Annual Meeting, *45*(2), 144–148. Los Angeles, CA: SAGE Publications.

Mouloua, M., Gilson, R., & Hancock, P. A. (2003). Human-centered design of unmanned aerial vehicles. *Ergonomics in Design*, *11*(1), 6–11.

Mouloua, M., Hancock, P., Jones, L., & Vincenzi, D. (2010). Automation in Aviation Systems: Issues and Considerations. In J. Wise, D. Garland, & D. V. Hopkin (Eds.), *Handbook of Aviation Human Factors*. Boca Raton: FL: CRC Press (Taylor & Francis Group) (Refereed, International).

Scott, M., & Doverspike, C. D. (2005). Training needs analysis and evaluation for new technologies through the use of problem-based inquiry. *Performance Improvement Quarterly*, *18*(1), 110–124.

Warm, J. S., Parasuraman, R., & Matthews, G. (2008). Vigilance requires hard mental work and is stressful. *Journal of Human Factors*, *50*(3),433–441.

3 Game-Based Small Team Training

A Guide to Implementing Adaptive Game-Based Simulation Training

Richard J. Simonson and Crystal M. Fausett

Literature on organizational effectiveness and team performance has illustrated the importance of teams in producing stronger and more effective outcomes (Salas et al., 2008a). Despite this known relationship, little emphasis has historically been placed on the importance of teamwork competencies during formal education and organizational training – particularly compared to taskwork competencies. The explanatory mechanism for this disproportionate training is how traditional training methodologies were developed and optimized for individual and technical knowledge and skill development. In contrast, teamwork is highly variable and based on soft skills (Gobeli, 2012). Additionally, barriers in adopting and integrating team training, identifying team training needs, and eliciting them are resource intensive (Cannon-Bowers & Salas, 1998) and are often only seen when the team engages in activities over some time (trust, psychological safety, etc. (Bohlander & McCarthy, 1996; Brasier et al., 2023)).

A common method used by organizations and researchers alike to mitigate these challenges is the use of Simulation-based training (SBT). SBT is a method of instruction that uses an interactive environment to replicate key features of "real-world" scenarios (Salas et al., 2009). Compared to traditional training methods, SBT presents a medium for teams to work together and elicit competencies and soft skills that only appear when teams engage in work. Further, SBT presents an opportunity to contextually train teams and tasks simultaneously by carefully considering cognitive and physical factors.

An extension of SBT, game-based ST (GBST), adds to the strengths of SBT and mitigates some of its limitations. GBST is the concept of using games as a modality to simulate the physical or cognitive characteristics and processes into SBT. The GBST method has been linked to increased trainee engagement compared to traditional and SBT methods (Sitzmann, 2011). This response increases motivation, learning, and transfer of the learner's training (Pellegrino & Scott, 2004; Prensky, 2003). Games have also become widely adopted across gaming and training industries, leading to cost-saving opportunities (Meliza et al., 2007). Additionally, the

DOI: 10.1201/9781032701639-3

format of a game as a training tool presents itself as a naturally adaptive framework that promotes stronger training outcomes, re-usability of training, and engagement from learners (Graafland et al., 2017; Ratwani et al., 2010). The adoption of GBST for team training by researchers and practitioners has significantly increased with technological capabilities and accessibility. However, developing and integrating games as a simulation-based tool requires careful planning and multi-disciplinary expertise to ensure success. Thus, the purpose of this chapter is threefold: (1) synthesize the theory and practice of team training research, (2) describe team training as implemented via GBST, and (3) provide guidance on how to utilize adaptive GBST strategies to improve small team training practices.

TEAM TRAINING

Teams are complex, fluid, and ever-evolving entities that form and perform toward a shared goal (Salas, Rosen et al., 2008). The characteristics of teams and their contributions to performance, coupled with their varying composition, significantly contribute to the infamous challenges faced in their research and development (Hamman, 2004; Sottilare et al., 2011). Subsequent efforts to decipher these characteristics and correlate them with performance and effectiveness have led to current frameworks of team competencies. A team's ability to work together effectively necessitates that teams must have proficiency in both the tasks they are completing, and the competencies required for effective teamwork. When evaluating the necessity for training, its success largely depends on these two main factors: the team's capability to collaborate effectively and their proficiency in accomplishing their designated tasks. A lack of skill in either teamwork or task execution can result in poor team performance. Conversely, if the training does not accurately address the specific area that needs improvement, whether it's teamwork or task-related skills, it may not be effective (Salas et al., 2008a). Cannon-Bowers et al. (1995) provide a clear explanation of the interplay between taskwork and teamwork. They refine existing frameworks related to both task and team dynamics, establishing a solid foundation for identifying the most effective content and methods for team training.

COMPETENCIES IN TEAM TRAINING

The basis of team training is centered around the measurement, assessment, and targeted improvement of team-based competencies. Competencies describe the characteristics an individual exhibits that contribute to the successful performance of a task. The overarching framework of team competencies used today originates in Bloom's taxonomy of learning domains (Engelhart et al., 1956), which breaks down learning competencies into knowledge, skills, and attitudes (KSAs). Under individual-level competencies, KSAs refer to one's knowledge necessary to perform a task (knowledge), their psychomotor capability to carry out the actions required to perform the task (skill), and their beliefs related to performing the task (attitude). The team framework of KSAs extends these definitions and integrates them into a hierarchical structure with both individual and team competencies and task- and team-related competencies. Under the teaming lens, knowledge represents the information and experience that team members have with respect to one another and the associated task. This competency stems from the research regarding shared mental models and their importance in teaming (Cannon-Bowers et al., 1995). Skills, within the teaming perspective, describe the observable behaviors exhibited by teams that contribute to team performance. Finally, attitudes are described as the beliefs that motivate teams to perform (Salas et al., 2008b).

Similar to individual training, teamwork KSAs are contextualized with the training objective and purpose. Cannon-Bowers et al. (1995) refined a model to establish and differentiate competencies based on context. They specify that knowledge, skills, or attitudes can fall into one of four categories: team-specific or generic competencies or task-specific or generic competencies. Team-specific competencies are KSAs that are affected by the characteristics of a team (e.g., trust that other team members believe in the team); in contrast, team-generic competencies are those whose KSAs are independent of specific team characteristics (e.g., communication and leadership skills). Task-specific competencies in teamwork are team behaviors that are associated with performance in certain tasks (e.g., shared knowledge of each other's task-specific roles); and task-generic team competencies describe those that apply regardless of the task (e.g., trust that teammates will complete their tasks).

While the categories of competencies are relatively constrained, the number of competencies associated with teamwork is, theoretically, endless. For example, in their literature review, Cannon-Bowers et al. (1995) identified 130 individual team-based skills alone. Further, when training within the context of task-based competencies, the number of KSAs included in training can quickly become overwhelming. Additionally, some competencies only form as the team works together, meaning that a competency that may warrant training will not yet have formed. Fortunately, multiple team-based competency models that target the most performance-related KSAs have been suggested. Notable models include Salas et al.'s (2005) proposal of a *big five* model of teamwork competencies that most contribute to a team's compelling performance, including team leadership, adaptability, mutual performance monitoring, backup behavior, and team orientation, which are dependent on the coordinating mechanisms of shared mental models, closed-loop communication, and mutual trust; or Salas et al.'s (2015) heuristics of teamwork which include cooperation, coordination, conflict, coaching, and communication, driven by three influencing conditions: context, composition, and culture.

Training Types and Context

Ensuring that team training is optimized for effectiveness and transfer to real work is an important consideration. Thus, choosing a strategy that most aligns with the training objective can significantly improve the training process. This section will explore a range of training strategies, focusing primarily on three key types: procedural training, cross-training, and adaptive training, with a special emphasis on adaptive training (Table 3.1).

Procedural training focuses on learner acquisition of specific sequences of actions (i.e., procedures) to accomplish a particular task or goal (Gorman et al., 2010).

TABLE 3.1
Summary of Training Types

Training Type	Description	Benefits	Limitations
Procedural Training	Focuses on learner acquisition of specific sequences of actions to accomplish tasks.	Useful in high-stakes settings to ensure standard protocol under stress.	Does not extend well to tasks with unforeseen variations.
Cross-Training	Involves training individuals on the responsibilities of their teammates to develop skills beyond their job functions.	Improves mutual understanding, shared mental models, coordination, and team performance.	May not benefit individual performance, less effective as team size and diversity increase.
Adaptive Training	Introduces controlled disruptions (perturbations) to enhance skill acquisition and performance.	Encourages flexibility and adaptation to new task challenges.	Requires sophisticated design and implementation to introduce and manage perturbations effectively.

Procedural training involves breaking complex tasks down into step-by-step procedures that can be followed. Often used in settings with high consequences for failure, procedural training helps remember a standard protocol under increased workload and stress levels, such as those in emergency response and military training (Hockey et al., 2007; Sauer et al., 2008). While this approach leads to skillful execution of the trained task, it does not extend well to tasks that involve unforeseen variations or disruptions, which are common in real-world situations (Ramakrishnan et al., 2017).

Cross-training involves training individuals on the responsibilities of their teammates. This allows individuals to learn and develop skills different from their job functions. Cross-training is beneficial for creating a mutual understanding of task-related work between team members and has been shown to improve shared mental models, coordination, and team performance (Marks et al., 2002). The cross-training method may not benefit individual performance, as individuals must learn several new roles. Further, cross-training does not scale as teams increase in size and diversity (Nikolaidis & Shah, 2013).

Adaptive training, sometimes known as perturbation training, involves carefully introducing controlled disruptions, or perturbations, into the learning process to enhance learners' skill acquisition and performance (Gorman et al., 2010). From the dynamic systems literature, a perturbation is the application of an outside force that briefly halts or otherwise disrupts a dynamic process, forcing that system to develop novel processes to adapt and return to the desired stable state (Gorman et al., 2010). In team training, perturbation is used to disrupt coordination procedures throughout the learning process intentionally. This compels the team to devise new methods to achieve their objectives. Perturbation training is "a human team-training strategy that requires team members to practice variations of a given task to help their team generalize to new variants of that task" (Ramakrishnan et al., 2017, p. 495). In contrast to training methods that involve varying the situation or objectives, perturbation training involves disrupting crucial coordination links while maintaining a crucial objective. Perturbation training aims to counteract habituation and procedural rigidity that can arise from cross-training and procedural training, respectively (Gorman et al., 2010). This approach enables teams to develop flexible interaction processes that can be applied to new and unfamiliar task conditions.

While perturbation training is one specific approach, adaptive training can also include other techniques. Adaptive training may involve dynamic simulations, scenario-based training, real-time adjustments based on learner performance, or personalized learning paths that adapt to the individual's needs. Another adaptive training technique is scaffolding. Scaffolding refers to offering support to students when required, gradually reducing that support as their competence and skills improve (Hogan & Pressley, 1997). Adaptive training systems have also been defined as "serious game-based systems whose goal is to engender communication opportunities for players to learn about their strengths and weaknesses, receive real-time in-game assessment feedback on their performance, and share diverse solutions and strategies during, between, and after a gameplay to update and adapt their understanding" (Raybourn, 2007, p. 206).

Scaffolding and Adaptive Training Techniques

Vygotsky & Cole's zone of proximal development (1978) refers to the distance between what a learner can do without help and what they can do with guidance (Raymond, 2000). Scaffolding, which refers to providing temporary support (scaffolds) to a learner to accomplish tasks, seeks individualized support based on a learner's zone of proximal development (Chang et al., 2002). As learners become more proficient, this support is removed or adapted based on the current level of performance. Examples of scaffolding as a training technique include guided instructions, hints, modeling of tasks, and questioning techniques. Other adaptive training techniques encompass several different approaches. Mastery learning is an approach in which learners are required to achieve a high standard of learning in one area before moving on to the next topic (Bloom, 1968). The differentiated instruction approach tailors teaching to individual needs, contrasting with traditional uniform methods that ignore students' unique needs (Suprayogi et al., 2017). Inquiry-based learning, however, promotes learning through questioning, exploration, and problem-solving (Friesen & Scott, 2013) (Table 3.2).

GAME-BASED SIMULATION TRAINING

SBT methods have been largely successful in their application to team training. However, they are still susceptible to limiting factors – most notably, prohibitive resource and monetary costs (Bell et al., 2008). Subsequently, researchers have sought to identify methods that can reduce costs and extend as well as increase the benefits of SBT (Sitzmann, 2011). One method that has gained recent popularity is the use of game-based training (GBT) methods, which use games or game elements to

TABLE 3.2
Summary of Adaptive Training Techniques

Adaptive Training Techniques	Purpose and Benefit
Scaffolding	Provides temporary support to learners, adjusting to their evolving competency levels, thereby enhancing skill acquisition.
Mastery Learning	Ensures learners achieve a high level of understanding and skill in one area before moving on, promoting deep knowledge and reducing skill gaps.
Differentiated Instruction	Tailors content and teaching methods to meet the diverse needs of learners, improving engagement and effectiveness.
Inquiry-Based Learning	Encourages learners to explore and ask questions, fostering critical thinking and adaptability.

enhance training and learning (Martens et al., 2008), and by extension, game-based approaches to simulation training.

GBT is a tool integrated into our everyday lives, from childhood to our professional interactions, in both virtual and physical mediums (Martens et al., 2008). While playing and engaging in games are intuitive to human nature, there exists discourse in the exact operational definition of a game and game-based simulation, and thus, a challenge in creating consistent interactions in GBT (Stenros, 2017). For this work, and in developing game-based simulation and training, we will use Browning's (2015) recommendation of essential game characteristics: a game must distract learners from the nature of the game (e.g., training) to promote engagement through play, the learner must be allowed to have choices that create dynamic events, there is no expectation of productivity tied to the real work conducted outside of the game, rules define the world and interactions that create obstacles the learner experiences in the game; additionally, the rules should create and awareness to the learner that the game is distinguishable from real work. The last point is essential to differentiate the context of this chapter concerning a game compared to a simulation. SBT relies on mimicking reality to the greatest extent possible, whereas a game's objective is to represent reality and engage players via gaming elements (Johnston & Whitehead, 2009; Narayanasamy et al., 2006).

Game Properties and Training

Games meant for training and education have historically been categorized as "serious games," but recent literature has found evidence that various game characteristics (not designed for the explicit purpose of training) are associated with increased learning (Pistono et al., 2021). The distinction between "serious games" and commercial games has become muddled; commercialized games sold primarily for entertainment purposes have quickly become serious games in educational and research settings (e.g., *Minecraft* (Microsoft, 2023; Nguyen & Rank, 2016); *Kerbal Space Program* (Rosenthal & Ratan, 2022; Take-Two Interactive Software, 2023);

Artemis Spaceship Bridge Simulator (Robertson, 2023; Simonson et al., 2021)), while initially serious games ranked have high in entertainment and enjoyability to entertainment-focused consumers (e.g., *America's Army* (Pellegrino & Scott, 2004; Shen et al., 2009; U.S. Army, 2002); also see (Hussain et al., 2008)). While seemingly at odds with traditional objectives of training, the use of commercialized games for training purposes suggests an important insight: effective training can be both educational and enjoyable. Learning and fun are not mutually exclusive in a well-developed training program.

Instead of classifying games as "serious games" or not, a more pragmatic approach to understanding games as a training medium is identifying the characteristics associated with learning processes and outcomes. A swathe of literature reviewed by Wilson et al. (2009) has provided, albeit not yet comprehensive, links between game characteristics and their ability to train various competencies. This research has also been extended to the team training domain. For example, Marlow et al.'s (2016) review of game attributes associated with team training identified eight game attributes: action language, assessment, conflict or challenge, environment, game fiction, human interaction, immersion, and rules or goals that mapped onto three teamwork competencies: coordination, communication, and cognition. Other literature has also identified specific games, both virtual and physical, that were correlated with increased proficiency in various teamwork competencies: Du Plooy and Parker (2020) utilized the physical-based *marshmallow game* for training teamwork, reporting the game's elements were effective in training and promoting team psychological safety; Martín-Hernández et al. (2021) noted a significant increase in intrinsic motivation, team engagement, team competence, and innovative behaviors from the physically based game called *the group to the rescue*; Peppen et al. (2022) successfully increased team situational awareness, decision making, communication, and resource management competencies via the web-based game *Team Up!* An important distinction to be made with respect to GBST is the distinction between gamified SBT and GBST. GBST necessitates that SBT becomes a game via adhering to the characteristics of what a game is (Browning, 2015), whereas a gamified SBT integrates game-like elements into a simulation (scoreboards, rewards, badges, etc.). While gamification does have beneficial properties in SBT, its properties are inherently different from those of GBST and will not be discussed in this chapter.

So far, we have discussed the theory and practice behind the complexities of teamwork and team training and have illustrated the recent adoption and benefits of the GBST approach. In summary, teamwork is a non-intuitive practice that takes purposeful and targeted training to improve. Traditional training methods are ill-suited to training team competencies due to their adaptive nature (Gorman et al., 2007). SBT mitigated many of the challenges associated with traditional team training but still suffers from various limitations that reduce adoption. The recent work and application of GBST has shown promising results in delivering the known benefits of SBT, while also addressing some of its weaknesses. However, GBST requires careful planning and development with interdisciplinary teams to ensure a successful and engaging training initiative. Therefore, the following section of this chapter will detail a literature-based guide on the best practices for developing GBST for small teams.

DEVELOPING ADAPTIVE GAME-BASED TEAM TRAINING

CONDUCT A TEAM NEEDS ASSESSMENT

The first step to any training initiative is understanding where the training should focus. Due to the complexity and emergent nature of many teamwork competencies within a team training context, it is imperative to accurately assess which competencies are lacking, otherwise known as a needs assessment. Conducting a needs assessment requires three steps: first, practitioners should determine which competencies to train. This process may include a search of the literature to identify which competencies are most associated with performance outcomes in the targeted domain and determine whether the training will need to include task- and team-specific or generic competencies. Next, determining whether training is an effective approach should be considered. Many factors can influence and negatively impact a team's performance, including individual interference (e.g., singular team members attitudes, behaviors, or competencies; poor team composition; poor or ineffective leadership), organizational influence (e.g., organizational culture), and external influence (e.g., environmental factors). If lacking competencies are affected by these or similar influences, then team training may not be the most effective approach. Finally, a training curriculum details what competencies will be trained, how they will be trained, and how their improvement will be compared to pre-established success parameters (Brown, 2002). A lack of a comprehensive understanding of the competencies before training may lead to ineffective training and poor staff outcomes (Cekada, 2011).

IDENTIFY SIMULATION CONTEXT – ESTABLISHING THE TRAINING DOMAIN

Next, one needs to set the simulation's context, or domain. Within GBST, the domain represents the game's characteristics and its scenario. This is a crucial step in the GBST process as this choice creates a boundary around what and how competencies can be trained (Rosen et al., 2008). Additionally, from a game-based perspective, the domain choice can mitigate or create barriers to training (Hussain et al., 2010). For example, suppose the training needs assessment identified that the team required communication training. In such a case, using a game that does not require information exchange may be far less effective, or possibly completely ineffective, at creating opportunities to elicit and train communication-related KSAs.

This step presents two possible paths: developing a custom game or using Commercial Off-The-Shelf (COTS) games. Custom-developed games provide extensive flexibility and versatility to practitioners, allowing for application to various purposes and training needs. However, developing a custom game is highly challenging, time-consuming, and requires development with interdisciplinary expertise in the targeted domain (e.g., teamwork), training, and game development (Hussain et al., 2010). COTS games, however, may reduce flexibility and versatility but are significantly more cost- and resource-efficient. However, COTS presents the challenge of finding the right game for the identified training needs. While research has linked certain gaming elements to trainable competencies, not all games perfectly summarize their elements, which can make finding the right game difficult (Simonson et al., 2023). Fortunately, GBT's growing popularity means a growing list of games, and their training effectiveness is available in the literature (Doherty et al., 2018; Hussain et al., 2008; Wilson et al., 2009).

SET LEARNING OBJECTIVES

Identifying competencies in the needs assessment is crucial to understanding what needs to be trained, but it does not provide information on how they will be trained. To gather this information, one needs to set the learning objects of the training. Learning objectives set the course for training by describing how one will measure the selected competencies and the point at which a trainee successfully or unsuccessfully gains proficiency in the competency. An important consideration in developing learning objectives, particularly in GBST, is determining how they will appear and be measured. For example, if the objective of a training initiative is to establish stronger communication skills, then understanding what facets of communication contribute to team effectiveness and performance is important. Identifying measurable competencies is completed by establishing the training's targeted KSAs.

SET THE TRAINING KSAs

Following the guidance of Salas et al. (2015), the competencies associated with effective team performance are endless. However, the extant literature provides evidence of task- and team-specific and generic competencies correlated with high-performing teams. The extensive nature of teamwork means that hundreds of competencies and even more variations on those competencies are possible. Therefore, we recommend the use of prior literature and domain-specific training Subject Matter Experts (SMEs) to assist in picking and integrating KSAs.

INTEGRATING THE GAME FRAMEWORK

Integration of games into SBT is an essential aspect of the GBST development process whose complexity depends on various factors. One primary factor is whether the decision is made to develop a custom game (either via game development or using game development platforms) and integrate it into simulation training or rely on a COTS game to provide the framework. If the former route is chosen, carefully considering their formation is necessary; otherwise, the benefits of GBST may be nullified. In this case, we highly recommend forming an interdisciplinary team with experience in training, expertise in the targeted domain, and game design and development to ensure success. Prior research on the best practices and game characteristics for review is also available to help promote the chances of success (see Hussain et al., 2010; Salen & Zimmerman, 2003). COTS games present a simpler and more direct approach to this step as their stories and engaging properties are already developed. But, as discussed previously, finding a game that can integrate and elicit the training context, learning objectives, and KSAs (Doherty et al., 2018; Simonson et al., 2023), as well as a luminous attitude in the trainees (Salen & Zimmerman, 2003) can be challenging. Fortunately, the growing popularity of game-based learning has led to increased awareness and sharing of game-competency associations.

GAME ELEMENTS AND KSA ELICITATION

Once the KSAs to train are established, the next step is to determine how the trainees will elicit them. Within traditional SBT, practitioners are encouraged to create set objectives that can be used to trigger a KSA-related response from the trainees (Grossman et al., 2014). This technique also applies to GBST. When implementing this technique, each learning objective should have at least one trigger to ensure completeness of training. An example of the trigger and response method in GBST is illustrated by Ramachandran et al. (2016), who utilized a workload management mechanic within their game-based learning study, causing one teammate to become overloaded with tasks. This trigger was used to elicit the expected response of team cohesion via other teammates assisting the overloaded team member.

While team competencies may be associated with certain triggers based on the training design, the reality is that the teaming competencies exhibited outside of these triggers are equally important. Fortunately, game-based approaches, especially within adaptive training frameworks, offer multiple other techniques to elicit competencies, dynamic simulations, scenario-based training, real-time adjustments based on learner performance, or personalized learning paths that adapt to the individual's needs. For example, Simonson et al. (2021) utilized their chosen games' built-in difficulty settings to create an additional challenge for the training teams; this method led to changes in competency elicitation throughout the game rather than at specific points in the game. Special consideration for the elicitation mechanisms may also be necessary when developing, testing, and iterating on the training design and elicitation methods. Continuing from the game-difficulty example, relying on a game's difficulty can present unforeseen challenges and require additional precaution. First, the overall difficulty changes may not affect the difficulty associated with the interactions that trigger competencies (Hussain et al., 2010; Ratwani et al., 2010). Second, game difficulty can significantly affect the opportunities to elicit teamwork competencies; too little difficulty may reduce engagement, and too much difficulty may overwhelm teams and reduce their ability to work together (Marlow et al., 2016).

KSA ASSESSMENTS AND CONSIDERATIONS

Due to the nature of time- and event-varying competency elicitation, many GBT initiatives and studies also choose to observe the competencies as they are exhibited rather than at specified time event points. However, this requires raters to observe and record the event. Subsequently, the choice of data collection modality and method and the ability to ensure high reliability from multiple observers should be carefully considered. As teams can exhibit competency from both unobservable cognitions and observable behaviors, some research suggests measuring the targeted competency via multiple methods (Koh et al., 2014; M. Rosen et al., 2010) to capture the knowledge, behavioral, and affective components of the competency. For example, mutual trust, described as "the shared belief that team members will perform their roles and protect the interests of their teammates" (Salas et al., 2005, p. 561), can be observed by the open sharing of information or teammates openly admitting mistakes, but they can also be tested by determining the level of agreement of each team members belief that other members will complete their tasks. Test-based and survey-based methods are effective at measuring the cognitive side of competencies. As a note, survey-based methods are contingent on the validity and reliability of the survey used; prioritizing psychometrically validated surveys is recommended. To illustrate this, in the case of backup behaviors, one could test each member's ability to accurately assess team workload proportions or use a survey that gauges each member's belief that their team exhibits backup behaviors (Salas, Rosen et al., 2008). In contrast, observation-based assessments are effective at seeing the behavioral characteristics of a competency. From a backup behavior perspective, an observation tool might record the backup behaviors as they occur. As with survey-based methods, observation tools vary by measurement and will affect the precision with which the competency is measured.

From a human-rater perspective, checklists, frequency counts, and rating scales are commonly used, each with advantages and limitations. Checklists are easy and quick to use but only allow for the recording of dichotomous events (Was the behavior elicited? Was it a positive or negative behavior? etc.); frequency counts extend checklists by allowing one to determine the density of the behavior (e.g., number of times it was exhibited); whereas behavioral rating scales add a dimension of quality assessment. Prior literature on team training has successfully utilized these tools, with some arguing that behavioral rating scales are imperative as teamwork competencies may lie on a continuum of quality (Griggs, 2021). Computer-based assessments are also available for assessing competencies during training, which can reduce the needed resources and time for training. Examples include Deaton et al.'s (2007) Enhancing Performance With Improved Coordination (EPIC) tool, which aids reviewers in reducing the workload associated with team assessments; or Alozie et al.'s (2020) Multimodal Integrated Behavior Analysis (MIBA) tool, which automatically tracks trainee movements and tracks specific elements (mouth movements, gaze, facial expression, etc.).

Finally, it's important to consider the level at which the competency is assessed. Team-based competencies are based on the interaction of multiple members with one another and the group as a whole. Subsequently, depending on the context, some competencies can be measured at the individual level (e.g., an individual's teamwork KSA) or at the team level (e.g., each member's competency score aggregated). This decision is based on the context of the training and learning objectives and the theoretical structure of the construct. For example, Lee et al. (2018) studied conflict in teams and differentiated individual conflict as a member's conflict with another member and team-level conflict as the team's members acknowledgment of that conflict.

BUILT-IN ADAPTIVE FRAMEWORKS

Adaptive training constitutes that a training session can dynamically change based on the learner's performance or that sequential training sessions adapt to prior performances. An adaptive assessment support system, or using technology to personalize and target the assessment process for distinct learners, can integrate this into training by altering the presentation of materials and assessment to best suit the individual learner(s). The process involves tailoring questions' content, sequence, and difficulty level (generally based on a learner's responses and performance). However, tracking performance and competencies through adaptive training can be a daunting task. Various technologies exist to support these efforts, but the effectiveness of these technologies is predicated on the adaptive mechanisms built into the training (see Arnold et al.'s (2013) approach using storyboarding or Hussain et al.'s (2010) approach using objective mapping). When these mechanisms are developed, they can be utilized to guide when and how the training adapts manually, or via the use of automated Learning Management Systems (LMS).

While adaptive training frameworks are feasible with both human-based and computer-based or aided assessment methods, they benefit greatly from computer-aided methods due to their time and scaling potential. LMS tools use algorithms and data analytics to continuously analyze performance, track progress, and provide insights for learners and educators/instructors. This allows for developing a personalized learning experience for students by quickly determining specific areas of strength or weakness and pinpointing gaps in knowledge. Subsequently, calibrating difficulty based on the team's performance can occur rapidly and automatically, which is a crucial element in optimizing the effectiveness of team training (Salas & Burke, 2002).

CONCLUSION

The objective of this chapter was threefold: introduce and elucidate the purpose and motivations behind team training, describe the benefits of GBST within an adaptive training framework, and provide a guide and resources on developing a GBST initiative. Our review of team training literature elucidated the complexity behind teams and their work. We described the relationship behind the characteristics of teams (e.g., competencies) that contribute to improved teamwork and further described their specific and generic association with teams and their tasks. We also reviewed the game-based learning and training literature to summarize and contextualize how games use their unique characteristics to develop a luminous attitude in their players, which is the primary engagement factor in GBST. We further provided a summary of the literature on games and their innate elements that lead to competency training capabilities. Finally, we provided a guide on utilizing GBST and adaptive training frameworks in team training contexts and suggested that the curriculum should include (1) a team training needs assessment, (2) identification of learning objectives based on training needs, (3) associating competencies and their KSAs to the set learning objectives, (4) a determination to create a custom game environment or to use a COTS game to integrate into the SBT, (5) methods to elicit the KSAs, both from a game-element and simulation-based approach, (6) guidance on how each KSA will be assessed as the modality (e.g., human-based or computer-based) of assessment, and finally (7) considerations on how training will adapt to learner performance, as well as tools and frameworks to track said performance changes.

REFERENCES

Alozie, N. M., Dhamija, S., McBride, E., & Tamrakar, A. (2020). *Automated Collaboration Assessment Using Behavioral Analytics.* Nashville, TN: International Society of the Learning Sciences.

Arnold, S., Fujima, J., Jantke, K. P., Karsten, A., & Simeit, H. (2013). Game-Based Training of Executive Staff of Professional Disaster Management: Storyboarding Adaptivity of Game Play. *Proceedings of the 2013 International Conference on Advanced ICT.* 2013 International Conference on Advanced ICT, Hsinchu, Taiwan. https://doi.org/10.2991/icaicte.2013.14

Bell, B. S., Kanar, A. M., & Kozlowski, S. W. J. (2008). Current issues and future directions in simulation-based training in North America. *The International Journal of Human Resource Management, 19*(8), 1416–1434. https://doi.org/10.1080/09585190802200173

Bloom, B. S. (1968). Learning for mastery. Instruction and curriculum. Regional education laboratory for the Carolinas and Virginia, Topical Papers and Reprints, Number 1. *Evaluation Comment, 1*(2).

Bohlander, G. W., & McCarthy, K. (1996). How to get the most from team training. *National Productivity Review, 15*(4), 25–35. https://doi.org/10.1002/npr.4040150405

Brasier, A. R., Burnside, E. S., & Rolland, B. (2023). Competencies supporting high-performance translational teams: A review of the SciTS evidence base. *Journal of Clinical and Translational Science, 7*(1), e62. https://doi.org/10.1017/cts.2023.17

Brown, J. (2002). Training needs assessment: A must for developing an effective training program. *Public Personnel Management, 31*(4), 569–578. https://doi.org/10.1177/009102600203100412

Browning, H. (2015). Guidelines for Designing Effective Games as Clinical Interventions: Mechanics, Dynamics, Aesthetics, and Outcomes (MDAO) Framework. In D. Novák, B. Tulu, & H. Brendryen (Eds.), *Handbook of Research on Holistic Perspectives in Gamification for Clinical Practice* (pp. 105–130). Hershey, PA, USA: IGI Global.

Cannon-Bowers, J. A., & Salas, E. (1998). Team performance and training in complex environments: Recent findings from applied research. *Association for Psychological Science, 7*(3), 83–87.

Cannon-Bowers, J. A., Tannenbaum, S. I., Salas, E., & Catherine, V. E. (1995). Defining Competencies and Establishing Team Training Requirements. In R. A. Guzzo, & E. Salas (Eds.), *Team Effectiveness and Decision Making in Organizations*, 1st ed. (pp. 333–380). Mahwah, NJ: Jossey-Bass.

Cekada, T. L. (2011). Need training? Conducting an effective needs assessment. *Professional Safety, 56*(12), 28–34.

Chang, K.-E., Sung, Y.-T., & Chen, I.-D. (2002). The effect of concept mapping to enhance text comprehension and summarization. *The Journal of Experimental Education, 71*(1), 5–23. https://doi.org/10.1080/00220970209602054

Deaton, J. E., Bell, B., Fowlkes, J., Bowers, C., Jentsch, F., & Bell, M. A. (2007). Enhancing team training and performance with automated performance assessment tools. *The International Journal of Aviation Psychology, 17*(4), 317–331. https://doi.org/10.1080/10508410701527662

Doherty, S. M., Keebler, J. R., Davidson, S. S., Palmer, E. M., & Frederick, C. M. (2018). Recategorization of video game genres. *Proceedings of the Human Factors and Ergonomics Society Annual Meeting, 62*(1), 2099–2103. https://doi.org/10.1177/1541931218621473

Du Plooy, E. J., & Parker, H., & Graduate School of Business, University of Cape Town, Cape Town, South Africa. (2020). Psychological safety and team learning during a problem-solving game for staff at a South African hospital. *Global Health Innovation*, *3*(1), 1–12. https://doi.org/10.15641/ghi.v3i1.867

Engelhart, M. D., Furst, E. J., Hill, W. H., & Krathwohl, D. R. (1956). *Taxonomy of Educational Objectives, Handbook I: The Cognitive Domain* (B. S. Bloom, Ed.). New York, NY: David McKay Co. Inc.

Friesen, S., & Scott, D. (2013). *Inquiry-Based Learning: A Review of the Research Literature.* Alberta Ministry of Education. http://galileo.org/focus-on-inquiry-lit-review.pdf

Gobeli, C. L. (2012). *Critical Design Factors for Effective Teamwork Training in the Workplace: A Survey of Training Professionals in Oregon.* [Dissertation, Oregon State University]. https://ir.library.oregonstate.edu/concern/graduate_thesis_or_dissertations/h128nj021

Gorman, J. C., Cooke, N. J., & Amazeen, P. G. (2010). Training adaptive teams. *Human Factors: The Journal of the Human Factors and Ergonomics Society*, *52*(2), 295–307. https://doi.org/10.1177/0018720810371689

Gorman, J. C., Cooke, N. J., Winner, J. L., Duran, J. L., Pedersen, H. K., & Taylor, A. R. (2007). Knowledge training versus processes training: The effects of training protocol on team coordination and performance. *Proceedings of the Human Factors and Ergonomics Society 51st Annual Meeting*, 51(4), 382–386.

Graafland, M., Bemelman, W. A., & Schijven, M. P. (2017). Game-based training improves the surgeon's situational awareness in the operation room: A randomized controlled trial. *Surgical Endoscopy*, *31*(10), 4093–4101. https://doi.org/10.1007/s00464-017-5456-6

Griggs, A. (2021). It's Not Just a Game: Exploring the Effects of an Escape Room Team Building Intervention [Dissertation, Embry-Riddle Aeronautical University]. *Doctoral Dissertations and Master's Theses.* https://commons.erau.edu/edt/625

Grossman, R., Heyne, K., & Salas, E. (2014). Game- and Simulation-Based Approaches to Training. In Kraiger, J. Passmore, N. R. dos Santos, & S. Malvezzi (Eds.), *The Wiley Blackwell Handbook of the Psychology of Training, Development, and Performance Improvement* (pp. 205–223). Hoboken, NJ: Wiley Blackwell.

Hamman, W. R. (2004). The complexity of team training: What we have learned from aviation and its applications to medicine. *Quality and Safety in Health Care*, *13*(suppl_1), i72–i79. https://doi.org/10.1136/qshc.2004.009910

Hockey, G. R. J., Sauer, J., & Wastell, D. G. (2007). Adaptability of training in simulated process control: Knowledge- versus rule-based guidance under task changes and environmental stress. *Human Factors: The Journal of the Human Factors and Ergonomics Society*, *49*(1), 158–174. https://doi.org/10.1518/001872007779598000

Hogan, K., & Pressley, M. (1997). Scaffolding Scientific Competencies within Classroom Communities of Inquiry. In K. Hogan & M. Pressley (Eds.), *Scaffolding Student Learning: Instructional Approaches and Issues* (pp. 74–107). Cambridge, MA: Brookline Books.

Hussain, T., Feurzeig, W., Cannon-Bowers, J., Coleman, S., Koenig, A., Lee, J., Menaker, E., Moffitt, K., Murphy, C., Pounds, K., Roberts, B., Seip, J., Wainess, R., Cannon-Bowers, J., & Bowers, C. (2010). Development of Game-Based Training Systems: Lessons Learned in an Inter-Disciplinary Field in the Making. In *Serious Game Design and Development: Technologies for Training and Learning* (pp. 47–80). Hershey, PA, USA: IGI Global.

Hussain, T., Weil, S. A., Brunye, T., Sidman, J., Ferguson, W., & Alexander, A. L. (2008). Eliciting and evaluating teamwork within a multi-player game-based training environment. Computer Games and Team and Individual Learning.

Johnston, H., & Whitehead, A. (2009). Distinguishing games, serious games, and training simulators on the basis of intent. *Proceedings of the 2009 Conference on Future Play on @ GDC Canada*, 9–10. https://doi.org/10.1145/1639601.1639607

Koh, E., Hong, H., & Seah, J. (2014). *An analytic frame and multi-method approach to measure teamwork competency*. 264–266. https://ieeexplore.ieee.org/stamp/stamp.jsp?arnumber=6901454&casa_token=uht8x6ZzFN4AAAAA:Fnuc63qF-DqiwY0cF-gxX0v0yBJsoCDzEVtdMRTqJBiqeFFUiTsx3MVhbs210A0vFK8wHSBV0SGE

Lee, S., Kwon, S., Shin, S. J., Kim, M., & Park, I.-J. (2018). How team-level and individual-level conflict influences team commitment: A multilevel investigation. *Frontiers in Psychology*, *8*. https://www.frontiersin.org/articles/10.3389/fpsyg.2017.02365

Marks, M. A., Sabella, M. J., Burke, C. S., & Zaccaro, S. J. (2002). The impact of cross-training on team effectiveness. *Journal of Applied Psychology*, *87*(1), 3–13. https://doi.org/10.1037/0021-9010.87.1.3

Marlow, S. L., Salas, E., Landon, L. B., & Presnell, B. (2016). Eliciting teamwork with game attributes: A systematic review and research agenda. *Computers in Human Behavior*, *55*, 413–423. https://doi.org/10.1016/j.chb.2015.09.028

Martens, A., Diener, H., & Malo, S. (2008). Game-Based Learning with Computers – Learning, Simulations, and Games. In Z. Pan, A. D. Cheok, W. Müller, & A. El Rhalibi (Eds.), *Transactions on Edutainment I. Lecture Notes in Computer Science* (Vol. 5080, pp. 172–190). Berlin Heidelberg: Springer. https://doi.org/10.1007/978-3-540-69744-2_15

Martín-Hernández, P., Gil-Lacruz, M., Gil-Lacruz, A. I., Azkue-Beteta, J. L., Lira, E. M., & Cantarero, L. (2021). Fostering university students' engagement in teamwork and innovation behaviors through game-based learning (GBL). *Sustainability*, *13*(24), 13573. https://doi.org/10.3390/su132413573

Meliza, L. L., Goldberg, S., & Lampton, D. R. (2007). *After Action Review in Simulation-Based Training* (RTO-TR-HFM-121-Part-II). U.S. Army Research Institute for the Behavioral and Social Sciences. https://apps.dtic.mil/sti/citations/ADA474305

Microsoft. (2023). *Minecraft* [Computer software]. Microsoft. https://www.minecraft.net/en-us

Narayanasamy, V., Wong, K. W., Fung, C. C., & Rai, S. (2006). Distinguishing games and simulation games from simulators. *Computers in Entertainment*, *4*(2), 9–es. https://doi.org/10.1145/1129006.1129021

Nguyen, A., & Rank, S. (2016). Studying the Impact of Spatial Involvement on Training Mental Rotation with Minecraft. *Proceedings of the 2016 CHI Conference Extended Abstracts on Human Factors in Computing Systems*, 1966–1972. https://doi.org/10.1145/2851581.2892423

Nikolaidis, S., & Shah, J. (2013). Human-robot cross-training: Computational formulation, modeling and evaluation of a human team training strategy. *2013 8th ACM/IEEE International Conference on Human-Robot Interaction (HRI)*, 33–40. https://doi.org/10.1109/HRI.2013.6483499

Pellegrino, J., & Scott, A. (2004). *The Transition from Simulation to Game-Based Learning*. Interservice/Industry Training, Simulation, and Education Conference, Orlando, FL.

Peppen, L., van, Faber, T. J. E., Erasmus, V., & Dankbaar, M. E. W. (2022). Teamwork training with a multiplayer game in health care: Content analysis of the teamwork principles applied. *JMIR Serious Games*, *10*(4), e38009. https://doi.org/10.2196/38009

Pistono, A. M. A. D. A., Santos, A. M. P., & Baptista, R. J. V. (2021). A review of adaptable serious games applied to professional training. *Journal of Digital Media & Interaction*, *4*(11), 60–85. https://doi.org/10.34624/JDMI.V4I11.26419

Prensky, M. (2003). Digital game-based learning. *Computers in Entertainment*, *1*(1), 21. https://doi.org/10.1145/950566.950596

Ramachandran, S., Presnell, B., & Richards, R. (2016). Serious games for team training and knowledge retention for long-duration space missions. *2016 IEEE Aerospace Conference*, 1–11. https://doi.org/10.1109/AERO.2016.7500503

Ramakrishnan, R., Zhang, C., & Shah, J. (2017). Perturbation training for human-robot teams. *Journal of Artificial Intelligence Research*, *59*, 495–541. https://doi.org/10.1613/jair.5390

Ratwani, K. L., Orvis, K. L., & Knerr, B. (2010). *An Evaluation of Game-based Training Effectiveness: Context Matters.* 1–10.

Raybourn, E. M. (2007). Applying simulation experience design methods to creating serious game-based adaptive training systems. *Interacting with Computers, 19*(2), 206–214. https://doi.org/10.1016/j.intcom.2006.08.001

Raymond, E. (2000). *Cognitive Characteristics. Learners with Mild Disabilities.* Boston, MA: Allyn & Bacon, A Pearson Education Company.

Robertson, T. (2023). *Artemis Spaceship Bridge Simulator* [Computer software]. Incandescent Workshop LLC.

Rosen, M., Weaver, S., Lazzara, E., Salas, E., Wu, T., Silvestri, S., Schiebel, N., Almeida, S., & King, H. (2010). Tools for evaluating team performance in simulation-based training. *Journal of Emergencies, Trauma, and Shock, 3*(4), 353. https://doi.org/10.4103/0974-2700.70746

Rosen, M. A., Salas, E., Silvestri, S., Wu, T. S., & Lazzara, E. H. (2008). A measurement tool for simulation-based training in emergency medicine: The simulation module for assessment of resident targeted event responses (SMARTER) approach. *Simulation in Healthcare: The Journal of the Society for Simulation in Healthcare, 3*(3), 170–179. https://doi.org/10.1097/SIH.0b013e318173038d

Rosenthal, S., & Ratan, R. A. (2022). Balancing learning and enjoyment in serious games: Kerbal Space Program and the communication mediation model. *Computers & Education, 182*, 104480. https://doi.org/10.1016/j.compedu.2022.104480

Salas, E., & Burke, C. S. (2002). Simulation for training is effective when…. *BMJ Quality & Safety, 11*(2), 119–120. https://doi.org/10.1136/qhc.11.2.119

Salas, E., DiazGranados, D., Klein, C., Burke, C. S., Stagl, K. C., Goodwin, G. F., & Halpin, S. M. (2008a). Does team training improve team performance? A meta-analysis. *Human Factors, 50*(6), 903–933. https://doi.org/10.1518/001872008X375009

Salas, E., Rosen, M. E., Shawn, B., & Goodwin, G. F. (2008b). The Wisdom of Collectives in Organizations: An Update of the Teamwork Competencies. In E. Salas, G. F. Goodwin, & B. Shawn (Eds.), *Team Effectiveness in Complex Organizations*, 1st ed. (pp. 73–114). New York, NY: Routledge.

Salas, E., Shuffler, M. L., Thayer, A. L., Bedwell, W. L., & Lazzara, E. H. (2015). Understanding and improving teamwork in organizations: A scientifically based practical guide. *Human Resource Management, 54*(4), 599–622. https://doi.org/10.1002/hrm.21628

Salas, E., Sims, D. E., & Burke, C. S. (2005). Is there a "Big Five" in teamwork? *Small Group Research, 36*(5), 555–599. https://doi.org/10.1177/1046496405277134

Salas, E., Wildman, J. L., & Piccolo, R. F. (2009). Using simulation-based training to enhance management education. *Academy of Management Learning & Education, 8*(4), 559–573.

Salen, S., & Zimmerman, E. (2003). *Rules of Play: Game Design Fundamentals.* Cambridge, MA: MIT Press.

Sauer, J., Burkolter, D., Kluge, A., Ritzmann, S., & Schüler, K. (2008). The effects of heuristic rule training on operator performance in a simulated process control environment. *Ergonomics, 51*(7), 953–967. https://doi.org/10.1080/00140130801915238

Shen, C., Wang, H., & Ute, R. (2009). Serious Games and Seriously Fun Games: Can They Be One and the Same? In R. Ute, M. Cody, & P. Vorderer (Eds.), *Serious Games*, 1st ed. (pp. 49–60). New York, NY: Routledge.

Simonson, R. J., Keebler, J. R., & Doherty, S. M. (2023). The need for recategorized video game labels: A quantitative approach. *Game Studies, 23*(1). https://gamestudies.org/2301/articles/simonson_keebler_doherty

Simonson, R. J., Keebler, J. R., Wallace, R. J., & Griggs, A. C. (2021). An investigation of team inputs, processes, and emergent states on performance in a spaceship bridge simulation. *Proceedings of the Human Factors and Ergonomics Society Annual Meeting, 65*(1), 1475–1479. https://doi.org/10.1177/1071181321651115

Sitzmann, T. (2011). A meta-analytic examination of the instructional effectiveness of computer-based simulation games. *Personnel Psychology, 64*(2), 489–528. https://doi.org/10.1111/j.1744-6570.2011.01190.x

Sottilare, R. A., Holden, H. K., Brawner, K. W., & Goldberg, B. S. (2011). *Challenges and Emerging Concepts in the Development of Adaptive, Computer-based Tutoring Systems for Team Training.* Interservice/Industry Training, Simulation, and Education Conference, Orlando, FL.

Stenros, J. (2017). The game definition game: A review. *Games and Culture, 12*(6), 499–520. https://doi.org/10.1177/1555412016655679

Suprayogi, M. N., Valcke, M., & Godwin, R. (2017). Teachers and their implementation of differentiated instruction in the classroom. *Teaching and Teacher Education, 67,* 291–301. https://doi.org/10.1016/j.tate.2017.06.020

Take-Two Interactive Software. (2023). Kerbal Space Program [Computer Software]. Take-Two Interactive Software.

U.S. Army. (2002). *America's Army* [Computer software]. U.S. Army.

Vygotsky, L. S., & Cole, M. (1978). *Mind in society: Development of higher psychological processes.* Cambridge, MA: Harvard University Press.

Wilson, K. A., Bedwell, W. L., Lazzara, E. H., Salas, E., Burke, C. S., Estock, J. L., Orvis, K. L., & Conkey, C. (2009). Relationships between game attributes and learning outcomes: Review and research proposals. *Simulation & Gaming, 40*(2), 217–266. https://doi.org/10.1177/1046878108321866

4 Game-Based Tools for Highly Automated Work

Trends, Challenges, and Opportunities

Alejandro Arca, James C. Ferraro, and Phillip M. Mangos

INTRODUCTION

The nature of work for many occupations is changing, evolving to adopt or develop new technologies to automate tasks previously performed manually. As automated systems and artificial intelligence (AI) proliferate the workspace, talent organizations must seek out the right skills and attitudes relating to working with automation (Brynjolfsson & McAfee, 2014). According to the U.S. Bureau of Labor Statistics (2024), while the federal government remains a critical employer in specialized and highly automated domains such as cryptography, cybersecurity, and intelligence analysis, private sector opportunities have experienced substantial expansion. For example, labor market projections indicate robust growth in information security professions. The U.S Bureau of Labor Statistics (2024) forecasts a 33% increase in information security analyst positions from 2023 to 2033, significantly outpacing the average growth rate for other occupations. This projection translates to approximately 17,300 annual job openings throughout the decade. Additionally, the cybersecurity workforce reached 5.5 million professionals globally in 2023, as documented by the ISC2 Global Workforce Study (ISC2, 2023). The workforce recruitment and training landscape in high-tech fields, including software development, cybersecurity, and military domains, reflects these significant technological and strategic shifts.

This chapter will investigate trends, challenges, and opportunities associated with game-based assessments for recruiting, assessment, and training in highly automated work environments. We will examine the current state of automation and AI in the workplace, the importance of effective human-automation interaction, and the potential of game-based assessments. This chapter will provide a comprehensive understanding of how organizations can better prepare their workforce for the future. Through a detailed analysis of current empirical research and a dive into trends from two independent market research studies, this chapter will offer evidence-based recommendations for integrating game-based assessments into the pre-hire and training processes, ultimately enhancing organizational efficiency and effectiveness.

DOI: 10.1201/9781032701639-4

The rapid evolution of workplace automation has profound implications for how organizations recruit, assess, and develop their workforce. As automated systems become more sophisticated and widespread, organizations must adapt their talent acquisition and development strategies to effectively select, train, and retain employees to work alongside these technologies. Understanding the current state of workforce automation provides a crucial context for examining how game-based tools can support these organizational objectives.

An Increasingly Automated Workforce

As organizations adapt to rapid technological advancement, the integration of automation and AI has emerged as a transformative force across industries. This technological evolution has particularly significant implications for workforce development, recruitment strategies, and skill requirements in both the military and civilian sectors. Recent analyses indicate substantial shifts in job roles and organizational structures as AI capabilities expand, warranting careful examination of these developments and their implications for future workforce planning.

The increasing role of automation and AI in the workplace is evident from recent survey data. A recent market research study from Adaptive Immersion (one of two that will be discussed within this chapter) with data collected from participants on Amazon's Mechanical Turk found that 81.90% of respondents reported utilizing automated systems (AI, robotics, etc.) to perform their job functions (Mangos & Ferraro, 2021). Looking more closely at the nature of their work, 51.43% of respondents reported currently using AI/ML in their day-to-day operations. While not a comment on *how* AI/ML is being applied in the workplace, this number further emphasizes the shift in how modern work is performed. This number is expected to grow, with projections indicating that 60.95% of respondents reported that they expect to use AI/ML within the next year, 73.33% within the next five years, and 68.57% within the next ten years (an admittedly harder projection to forecast), as seen in Figure 4.1.

These statistics underscore the pervasive integration of AI and automation in the workplace, highlighting the need for effective human-automation interaction. As AI and automation become more embedded in daily operations, the ability to work seamlessly with these technologies will be crucial for organizational success (Autor, 2015).

Challenges Interacting with AI and Automation

Effective human-automation interaction is critical to maximizing the benefits of AI and automation in the workplace while minimizing potential drawbacks. Parasuraman and Riley (1997) identified the possible misuse, disuse, and abuse of automation as significant concerns. Described frequently in human factors' literature, these three concepts represent the possible ways in which automated systems may be utilized to not result in optimal system performance.

- Misuse: The tendency to rely too heavily on the performance of automated systems or AI. Overestimating its capabilities and failing to effectively attend to the system and correct errors.

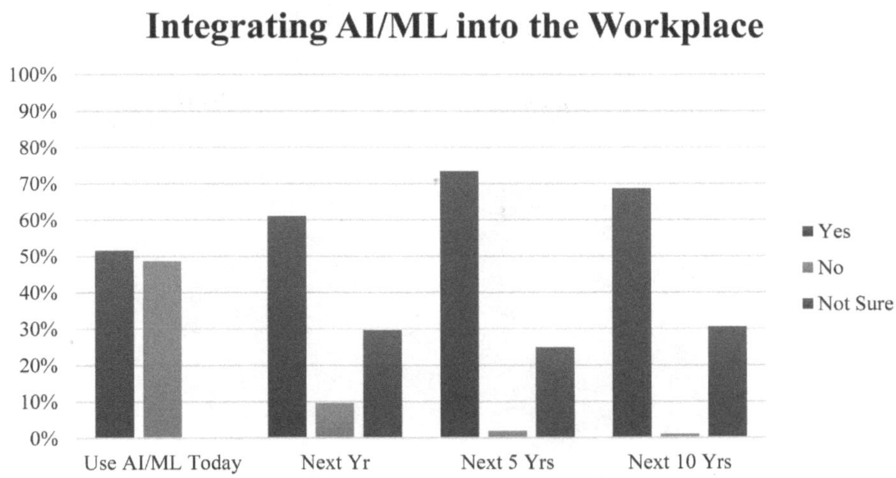

FIGURE 4.1 Market trends in organizations integrating AI/ML into day-to-day operations.

- Disuse: The tendency to not utilize automated systems or AI in the context in which they are intended. This results in a failure to see the full potential of the system.
- Abuse: Deploying automated systems or AI to take on tasks without considering the impact on the human or the system.

We believe these aspects of human-automation use, first addressed nearly 30 years ago, will continue to pervade human-machine systems as the technology evolves and expectations of the user become less predictable. Market research analyses suggest that this may be the case, with more than half of all respondents reporting having experienced or witnessed human performance issues when interacting with automation/AI (Mangos & Ferraro, 2021). The results of this market research study revealed interesting trends in the types of issues that were identified as most commonly experienced or witnessed in high-tech fields. With many of these tools still admittedly in their infancy relative to traditional methods of performing tasks, either manually or with simpler technologies, it is interesting to see that automated systems and AI tend to be underutilized in the workplace. Over one in three survey participants felt that co-workers tend to underestimate the capabilities of automation and/or AI.

Respondents were asked to identify some of the human factors and performance issues they have witnessed in their workplace. Specifically, they were asked whether they have seen problems related to situation awareness, workload management, overreliance on the technology, underutilization of the technology, or boredom with new hires to high-tech positions. These can all be considered symptoms (or direct examples) of automation misuse, disuse, or abuse. The results, shown graphically below, suggest a seemingly clear trend in how automation and AI are being used and adopted by new employees.

Over half of the participants have seen issues related to the division of labor between automation and themselves, balancing workload as they work in tandem

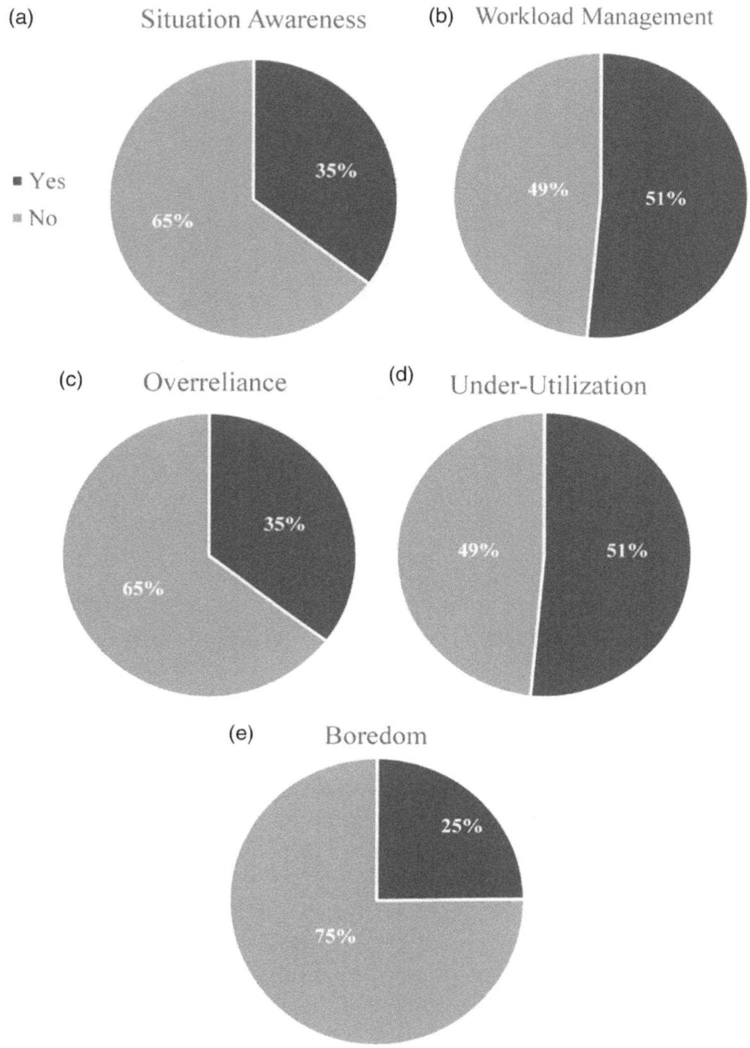

FIGURE 4.2 Market analysis of reported human factors issues interacting with AI and automation, including situation awareness (a), workload management (b), overreliance (c), under-utilization (d), and boredom (e).

with technology to complete tasks. Similarly, over half of the participants reported that automation and AI are underutilized in the workplace. Meanwhile, as it pertains to situation awareness, overreliance, and boredom, a much smaller group of participants reported witnessing these issues, as seen in Figure 4.2.

This general trend indicates that automated tools and AI are not being sufficiently utilized, leading to potential inefficiencies and missed opportunities (Endsley, 2017; Parasuraman & Riley, 1997). Over half of the respondents have observed issues related to the division of labor between automation and themselves, as well as challenges in balancing workloads. This underutilization of automation and AI in the workplace highlights the need for better training and assessment methods to ensure employees can effectively collaborate with these technologies (Hoff & Bashir, 2015).

The prevalence of automation misuse, disuse, and abuse in the workplace points to a critical need for more effective methods of evaluating and developing employees' capabilities to work with automated systems. Traditional assessment and training approaches often fall short in preparing workers for the complexities of human-automation interaction. Game-based assessment tools offer a promising solution to these challenges by providing immersive, realistic environments where candidates can demonstrate their ability to effectively collaborate with automated systems while organizations can evaluate their potential performance.

GAME-BASED ASSESSMENT AS A SOLUTION

Recruiting and assessing candidates for roles involving automated systems present unique challenges. Hiring managers aim to identify and prepare the most qualified candidates, often needing to sift through hundreds to thousands of resumes and personal statements to make these decisions. To accelerate this process, automated systems and even AI have been integrated into the pre-hire assessment process, weeding out applicants who do not meet certain criteria in the eyes of the system. This may speed up the process, but it is a far from perfect method of identifying the most qualified candidates for a job. Organizations should constantly aim to improve their pre-hire assessment process, and there may be an opportunity to integrate game-based assessments to address some of the shortcomings that exist. Game-based assessments could offer a solution by providing a more interactive and engaging way to evaluate these skills (Shute & Ventura, 2013).

Game-based assessments and serious games represent a promising approach to addressing the challenges of recruiting and training for roles involving AI and automated systems. Despite the potential benefits, market research revealed that 63.11% of workers reported not using game-based assessments in their current processes (Mangos & Ferraro, 2021). This presents an opportunity for organizations to adopt innovative assessment methods that can better evaluate candidates' abilities to work with AI and automation and effectively recruit the highest level of talent. Game-based assessments can simulate real-world scenarios, allowing candidates to demonstrate their skills in a controlled environment. This approach can provide valuable insights into how individuals interact with automated systems, identify areas for improvement, and ensure that new hires are well-equipped to handle the demands of their roles (Gee, 2003).

Research strongly supports an increased usage of entertainment video games for training and recruiting purposes. For instance, the game America's Army, developed in 2002, has since been played by more than 15 million individuals and evaluated as one of the most effective recruitment tools for the Army (Edery & Mollick, 2008). A market research survey revealed that 30% of Americans aged 16–24 had a more positive impression of the Army because of the game, and the game had more impact than all other forms of Army advertising combined.

Attraction to Game-Based Tools

Adaptive Immersion completed an independent market research study to gain feedback on the potential for game-based tools for pre-hiring assessment from individuals representing potential recruits for NASA, DoD, IC, and STEM industry-related fields. This survey targeted individuals in the current labor market closely resembling the demographics recruits for these industries. The purpose of the market analysis survey was to measure the propensity of these individuals to engage in a game-based assessment process for hiring, on-the-job learning, and career advancement in the target career fields. The key findings of the market analysis survey are as follows:

- Overall, 82.5% report a strong propensity (M = 4.27 on a 1–5 scale) to apply to a job that includes a game component.
- Respondents with a strong background in STEM fields are willing to put more time (M = 3.74) into completing a game-based application than a traditional pen-and-paper application.
- Respondents with a strong background in STEM fields report strong agreement (M = 3.97) that a game-based application would be an effective recruiting method.
- Respondents with a strong background in STEM fields are more likely (M = 3.66) to persist on the application even when embedded game challenges became more difficult.

In an age where tech organizations are in competition to identify and recruit top talent, this information can prove to be invaluable to improving the capabilities of their workforce. An example of this method being applied took place in 2016, when the company Unilever revamped its hiring process, utilizing a series of neuroscience-based games to identify quality candidates (Feloni, 2017; Wilson et al., 2018). Candidates were measured on their ability to perform well in these games before advancing to a video-based interview with an automated system. After one year of transitioning to this new hiring method, incorporating the game-based assessment as an initial barrier to the interview process, candidates began applying from over 2,500 universities, up from almost 850. Additionally, the acceptance rate of offers extended to applicants rose from 64% to 82%, suggesting that applicants left the pre-hire process with a positive impression of the company and a desire to work there.

Deploying game-based assessments for training, recruiting, and assessing a pre-hire's cognitive abilities has several advantages. In the pre-hire process, they tend to be far less intimidating than traditionally administered ability tests. This can reduce test anxiety and better capture applicant performance. It also makes the application process less cumbersome, creating a positive mental image of the organization in the eyes of the applicant. In both pre-hire and the training processes, a game-based element can help drive engagement and immersion in the task, enhancing motivation and even encouraging applicants to continue in the application process.

Training for highly automated jobs can also benefit from the use of game-based methods of evaluating and measuring performance. Research has found that game-based training can be applied in various STEM domains, such as healthcare and transportation, to improve performance in highly specialized tasks. In 2019, researchers examined the use of virtual reality (VR) serious games for training drivers' interactions with highly automated vehicles (Ebnali et al., 2020. The use of the game-based VR trainer resulted in faster reaction times in takeover scenarios and fewer overall collisions in future scenarios. Notably, additional results also suggested improvements in trust and acceptance of the automated driving technology. Similarly, there have been observed differences in overall self-efficacy in performing essential healthcare tasks such as chemotherapy preparation using game-based training methods (Garnier et al., 2024). This provides evidence that game-based training with highly automated systems can assist in changing attitudes toward the system or about the trainee's performance and potentially support more rapid adoption of new technologies.

When considered within the context provided earlier in this chapter, it is suggested that more common performance issues with these systems, including underutilization and applying game-based training for new hires, may help in getting the most out of the available automated systems. It can identify and even influence attitudes toward highly automated systems (e.g., self-driving cars) and help enhance performance in human-machine teaming tasks. Adaptive Immersion's market research study revealed that while over 60% of respondents stated their organization does not currently utilize game-based methods of evaluating performance, nearly half (47.57%) indicated that they would be likely to use one in the future. Almost one in three participants reported that they felt a game-based assessment tool aimed at evaluating human-automation interaction would be useful.

The strong attraction to game-based assessment tools among potential candidates, particularly in STEM fields, has led organizations to examine how these tools compare to traditional evaluation methods. Understanding the relative strengths and limitations of game-based assessments versus conventional approaches is crucial for organizations seeking to optimize their talent acquisition processes. Market research provides valuable insights into how different stakeholders perceive the comparative value of these assessment methods.

Perceived Comparative Value for Candidate Evaluation

The benefits of game-based pre-hire and training assessment tools described above can be applied as standalone assessments or, as Unilever deployed them, as a barrier somewhere within the hiring process. Game-based methods of evaluating a candidate for a job may elevate the quality of that evaluation. Market research has shown that a majority of surveyed workers feel that a game-based assessment may be more valuable to the hiring process than other traditional methods (Mangos & Ferraro, 2021). Summarized below are several of the more commonly used and traditional pre-hire talent assessment and acquisition tools.

- *Prior Experience/Knowledge*: Evaluations based on years of experience in a particular field or relevant experience in a related field.

- *Candidate Interview*: A face-to-face or virtual conversation with the applicant wherein they are asked job-specific questions by a hiring manager or employee in their desired position.
- *Personality Assessment*: Evaluate the overall fit of an applicant into an organization's culture and assess how likely they are to stick around long-term.
- *Cognitive Skill Assessment*: Evaluate the underlying abilities that can reveal performance in key performance areas such as decision-making and judgment.

Market analysis suggests that the perception of game-based assessments is more positive generally than many of these alternatives (Mangos & Ferraro, 2021). For example, 61.76% of respondents reported that a game-based assessment would be more valuable to their hiring process than a personality assessment. Additionally, 56.31% of respondents reported that a game-based assessment would be more valuable to their hiring process than based solely on previous knowledge or experience. Finally, nearly two-thirds (66.02%) of respondents reported that a game-based assessment would be more valuable to their hiring process than a non-game-based cognitive skill assessment alternative. The positive perception, both in engaging new hires and in evaluating applicants, suggests that game-based tools should be more prevalent than they seem to be. Empirical research into how these tools are being used and how effective they are indicates that researchers are paying more attention to this emerging trend.

Research Trends in Game-Based Assessment for Highly Automated Domains

In addition to examining market trends in how game-based tools are used for prehire assessment and training, Adaptive Immersion performed an empirical trend analysis of published research in game-based assessment, selection, and recruiting. Utilizing the Web of Science online database to identify relevant articles published between 2000 and 2024, the search criteria were designed to capture publications that included one of the following terms in either the title, abstract, or keywords: "Game-based assessment," "Game-based selection," or "Game-based recruiting." This approach facilitated a comprehensive examination of research trends in game-based assessment, selection, and recruiting spanning nearly two and a half decades.

By focusing on peer-reviewed articles, conference proceedings, and scholarly book chapters, the study aimed to maintain a high standard of academic rigor. The inclusion of early access articles allowed for the consideration of cutting-edge research that may not yet have been formally published but has undergone initial peer review. The trend analysis revealed significant growth and evolution in research development in game-based assessment, selection, and recruiting from 2000 to 2024. The total number of publications meeting the criteria was 415, encompassing a mix of articles, proceeding papers, review articles, book chapters, and early access articles.

Temporal trends demonstrated a clear upward trajectory in publication numbers. The 2000s decade from 2000 to 2009 saw 16 publications, followed by a substantial increase to 203 publications in the 2010s decade from 2010 to 2019. The field maintained strong momentum into the early 2020s, with 196 publications from 2020 to 2024, as seen in Figure 4.3.

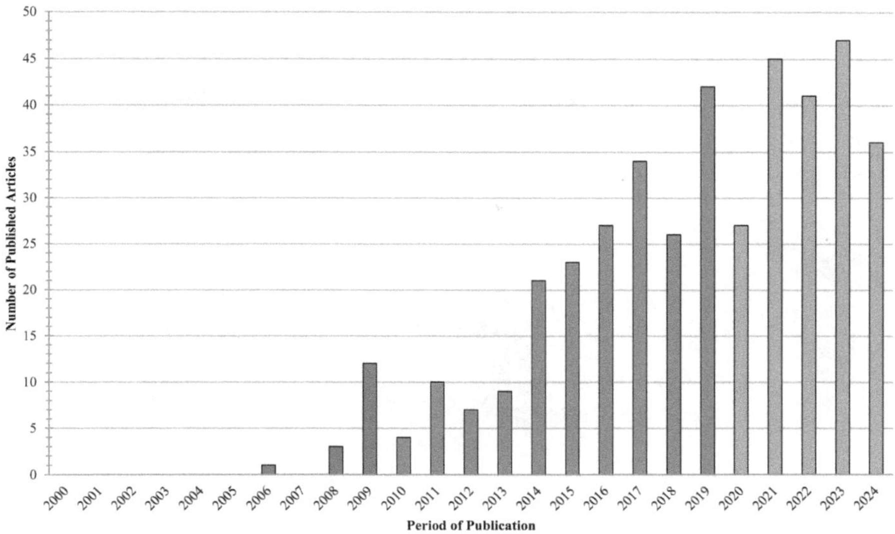

FIGURE 4.3 Annual trends in research into game-based tools for selection, assessment, and recruiting.

Furthermore, the analysis of publication trends reveals a remarkable acceleration in research output within the domains of game-based assessment and selection. This current decade, despite being only in its initial phase, has already produced a volume of research nearly equivalent to that of the entire preceding decade. This phenomenon is elucidated in Figure 4.4, which illustrates the mean annual publication rate across three distinct periods of publication.

The application domains represented in the publications underscore the interdisciplinary nature of game-based assessment and selection research. Education Research led with 188 publications (45.4%), followed by Computer Science with 150 publications (36.2%). Psychology and Engineering also showed significant contributions, with 59 (14.3%) and 56 (13.5%) publications, respectively. Other notable domains included Telecommunications, Business Economics, Social Sciences, Transportation, and various health-related fields. An interesting trend revealed that funding for this research came largely from military and governmental organizations.

These entities supported 195 publications (47.1%), followed by academia funding 154 publications (37.2%). Non-profit organizations contributed to 55 publications (13.3%), while industry funding was limited to 11 publications (2.7%). In conclusion, this trend analysis reveals a rapidly growing and interdisciplinary field of research in game-based assessment and selection. The exponential increase in publications is evident, with the average annual output doubling each decade from 1.6 publications per year in the 2000s to 20.3 in the 2010s and reaching 39.0 in the early 2020s, as seen in Figure 4.5. This trajectory suggests a continued expansion of the field.

The substantial growth in research attention to game-based assessment, particularly in highly automated domains, reflects the increasing recognition of these tools'

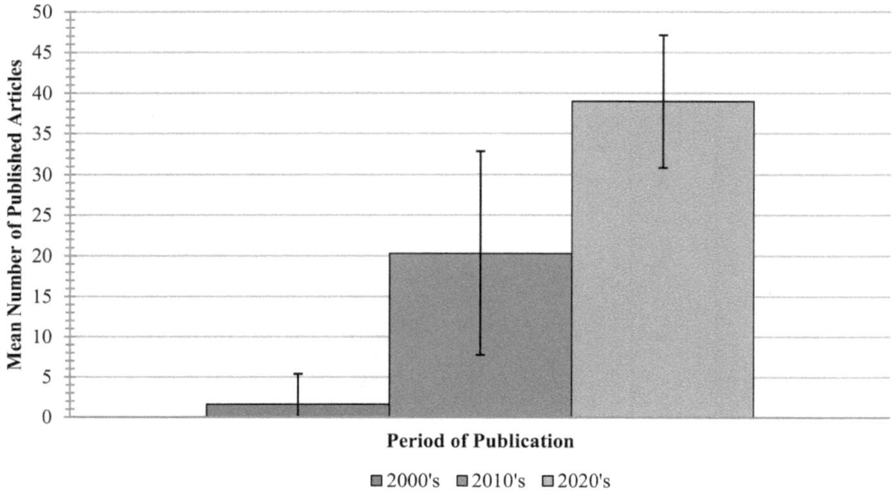

FIGURE 4.4 Average research publications per decade in game-based selection, assessment, and recruiting.

potential value for organizations. This convergence of empirical evidence and practical application provides a foundation for identifying key trends and implications for the future of workforce development in automated environments. By synthesizing insights from both research literature and market analysis, we can derive actionable recommendations for organizations seeking to leverage game-based tools in their talent management strategies.

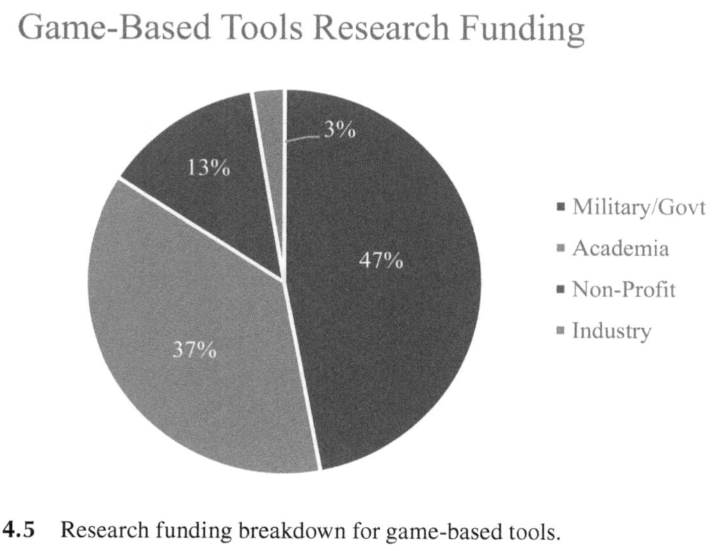

FIGURE 4.5 Research funding breakdown for game-based tools.

KEY TRENDS AND TAKEAWAYS

This chapter has summarized the current state of game-based tools used for pre-hire assessment and training, highlighting its perception as an effective method for delivering assessment material and identifying opportunities in STEM fields to apply the technology. The market research studies discussed above provide interesting insights into how new technologies, including AI, are being applied in the workplace. Despite the apparent proliferation of automated and AI features in the way work is performed (Mangos & Ferraro, 2021), the prevailing trend was that these systems are underutilized and their capabilities underestimated in the workplace (Mangos & Ferraro, 2021). Due to the still nascent capabilities of modern AI systems and automation (e.g., natural language processing and deep learning) in the grand scope of technology development, it is possible that a lack of exposure has engendered a lack of trust. There is likely a segment of the workforce that does not feel it necessary to utilize AI or new automated systems or features to do and/or assist with their work. Another segment does not want to appear ignorant of how to best utilize these automated systems and thus is hesitant to adopt them.

We have presented above examples of how game-based tools for recruiting, assessment, and training can help bring in the most qualified workers and even help modify their attitudes toward automation and AI (Ebnali et al., 2020; Feloni, 2017; Mangos et al., 2020; Wilson et al., 2018). There appears to be an opportunity to expand upon the applications for the game-based methods of identifying and training top talent, particularly in STEM fields (Mangos & Ferraro, 2021). Empirical research trends suggest that these tools are gaining more attention in the scientific community, and as evidence builds for their efficacy as recruiting and training methods, they may be adopted more commonly across industries. An emerging trend that may point to the direction of future research in this area is the number of funded projects sponsored by the government or military. Recruiting has long been a priority for the U.S. military, and presenting recruits with a more attractive platform for breaking into the armed forces may provide a boost to their numbers.

Gaming for Military and Commercial STEM Recruiting

The intersection of gaming and military recruitment represents a critical evolution in talent acquisition strategies. Recent data demonstrates the strategic value of gaming platforms for military and STEM recruitment, with approximately 75% of active-duty U.S. military personnel engaging with video games. This high engagement rate, particularly among younger service members, presents a compelling opportunity for recruitment initiatives.

In the civilian sector, gaming engagement closely mirrors military participation rates. The Entertainment Software Association (2023) reports that 69% of Americans regularly participate in gaming activities, with first-person shooter (FPS) games dominating the market. These games comprise 42% of Steam's Platinum-level Top Sellers in 2023, exemplified by the Call of Duty franchise's continued success (Steam, 2023). The launch of Modern Warfare II and Warzone attracted over 25 million players within just five days, demonstrating the massive reach and influence of gaming platforms.

While the pioneering military recruitment game "America's Army" concluded its two-decade run in 2022, military branches have significantly evolved their digital recruitment strategies. The U.S. Navy now dedicates a substantial portion of its marketing budget to esports initiatives, investing up to $4.3 million annually in gaming-related recruitment efforts. The Army has strengthened its gaming presence by establishing the Army Gaming League, which connects service members through competitive gaming environments. Similarly, the Air Force has expanded its reach by sponsoring major esports tournaments and maintaining dedicated streaming channels. Even the Space Force launched its "Space Force Gaming" initiative in 2022 as an innovative approach to fostering recruitment and community building.

The gaming-based recruitment model has expanded beyond military applications, finding significant traction in commercial STEM sectors. Major technology companies have adopted gaming principles by hosting immersive coding competitions and hackathons that simulate real-world problem-solving scenarios. Defense contractors have integrated simulation-based assessment games into their hiring processes for technical roles, allowing candidates to demonstrate their skills in realistic environments. Cybersecurity firms have particularly embraced this approach, implementing capture-the-flag (CTF) competitions that effectively identify and evaluate talented security professionals.

Looking toward the future, several emerging trends are reshaping the landscape of game-based recruitment. Organizations are increasingly willing to incorporate AI-powered assessment metrics within gaming environments, enabling more sophisticated evaluation of candidates' capabilities. VR training simulations are becoming more prevalent for high-risk operations, allowing organizations safety when assessing performance in various challenging scenarios. The development of cross-platform recruitment games has enabled organizations to evaluate both technical proficiency and essential soft skills simultaneously. Additionally, organizations are implementing gamified continuous learning platforms that support ongoing workforce development and skill enhancement.

These developments signal a significant shift toward more sophisticated, data-driven approaches to game-based recruitment in military and commercial STEM sectors. Organizations increasingly recognize gaming platforms as valid assessment tools that effectively evaluate candidates' problem-solving abilities, team coordination, and technical aptitude in realistic scenarios. As technology evolves and gaming platforms become more sophisticated, the integration of game-based assessment tools in recruitment and training processes is likely to become even more prevalent across industries.

REFERENCES

Autor, D. H. (2015). Why are there still so many jobs? The history and future of workplace automation. *Journal of Economic Perspectives*, *29*(3), 3–30.

Brynjolfsson, E., & McAfee, A. (2014). *The Second Machine Age: Work, Progress, and Prosperity in a Time of Brilliant Technologies*. New York, NY: WW Norton & Company.

Ebnali, M., Kian, C., & Ebnali-Heidari, M., & Mazloumi, A. (2020). User experience in immersive VR-based serious game: an application in highly automated driving training. In *Advances in Human Factors of Transportation: Proceedings of the AHFE 2019 International Conference on Human Factors in Transportation, July 24–28, 2019, Washington DC, USA 10* (pp. 133–144). New York, NY: Springer International Publishing.

Edery, D., & Mollick, E. (2008). *Changing the game: How video games are transforming the future of business*. FT Press. Upper Saddle River, New Jersey.

Entertainment Software Association. (2023). *2023 essential facts about the U.S. video game industry*. https://www.theesa.com/resources/essential-facts-about-the-us-video-game-industry/2023-2/

Endsley, M. R. (2017). From here to autonomy: Lessons learned from human–automation research. *Human Factors*, *59*(1), 5–27.

Feloni, R. (2017, June 28). Consumer-goods giant Unilever has been hiring employees using brain games and artificial intelligence - and it's a huge success. Retrieved 1/2/2025, from http://www.businessinsider.com/unilever-artificial-intelligence-hiring-process-2017-6

Garnier, A., Bonnabry, P., & Bouchoud, L. (2024). Game-based learning as training to use a chemotherapy preparation robot. *Journal of Oncology Pharmacy Practice*, *30*(4), 661–672.

Gee, J. P. (2003). What video games have to teach us about learning and literacy. *Computers in Entertainment (CIE)*, *1*(1), 20.

Hoff, K. A., & Bashir, M. (2015). Trust in automation: Integrating empirical evidence on factors that influence trust. *Human Factors*, *57*(3), 407–434.

ISC2. (2023). *Cybersecurity workforce study*. https://media.isc2.org/-/media/Project/ISC2/Main/Media/documents/research/ISC2_Cybersecurity_Workforce_Study_2023.pdf

Mangos, P. M., Boettcher, S., & Hulse, N. (2020). *Analysis of gaming preferences and market demographics for military-related serious games (Adaptive Immersion Technical Report)*.

Mangos, P. M., & Ferraro, J. C. (2021). *Market analysis: Automated and unmanned systems recruiting and selection practices (Adaptive Immersion Technical Report)*.

Parasuraman, R., & Riley, V. (1997). Humans and automation: Use, misuse, disuse, abuse. *Human Factors*, *39*(2), 230–253.

Shute, V., & Ventura, M. (2013). *Stealth assessment: Measuring and supporting learning in video games*. The MIT Press. Cambridge, MA.

Steam. (2023). *Best of 2023*. https://store.steampowered.com/sale/BestOf2023

U.S. Bureau of Labor Statistics. (2024, May). *Information security analysts. Occupational outlook handbook*. https://www.bls.gov/ooh/computer-and-information-technology/information-security-analysts.htm

Wilson, M., Kurzweil, M., & Alamuddin, R. (2018). Mapping the wild West of pre-hire assessment.

5 Artificial Intelligence Explainability
A Human Factors Approach

Gabriella M. Hancock, Laura M. Ornelas, Theresa Kessler, Tracy L. Sanders, and P. A. Hancock

INTRODUCTION

Artificial intelligence (AI) is a collective term encompassing broad-scale machine systems designed to simulate human cognitive processes in an attempt to produce decisions and actions on par or superior to those of human agents. Searle (1980) drew an important distinction between two key types of AI (weak and strong) that have critical philosophical impacts on the way such systems are designed, developed, and used. Weak AI envisions the machine as a powerful tool, one which naturally overcomes the resource-limited and structurally limited nature of humans to process information more effectively and efficiently for the benefit of the human decision-maker or agent. Strong AI, however, entails the generation and growth of computer programs that inherently understand said information, learn from it, and make their own decisions and actions based upon it. Strong AI consequently is said to have its own mind, consciousness, and cognitive states that emulate but are not equivalent to those of humans; it is its own unique entity. While few individuals have made credible claims as to the existence of strong AI, weak AI not only exists but has already prolifically permeated critical large-scale industries and applications throughout the world, including education (Nemorin et al., 2023), medicine (Kulkarni et al., 2020), finance (Chen et al., 2023), law (Reiling, 2020), and climate advocacy and action (Stein, 2020). More recently, generative AI (e.g., ChatGPT) large language models (LLMs) have become popular for their ability to use natural language processing to respond to natural language prompts and generate coherent text. Similarly, general adversarial (GAN) or diffusion models are able to generate images based on user prompts (Sætra, 2023).

AI is already making in-roads in terms of implementation in human performance measurement on the key fronts of interest: training, selection, and assessment, and generative AI promises to make these options more accessible (Budhwar et al., 2023). While humans excel at complex decision-making, AI has not yet reached the point of being capable and effective at making decisions based on ambiguous or unevenly weighted information, which is often required in both assessment and

 DOI: 10.1201/9781032701639-5

selection. In these cases, AI often provides support to a human for these critical decisions. As a result, explainability is necessary to be able to evaluate the AI's recommendations and ensure that it does not provide that support based on faulty information or use inappropriate biases (as has occurred in the past with serious ethical implications). Moreover, in cases wherein AI is making those final decisions, it is even more important that the reasoning behind said decisions can be reviewed and evaluated appropriately.

Hence, AI – and particularly those systems designed for human performance assessment – must be designed in a human user-centered manner. The AI is not autonomous and does not decide what goals it wants to achieve or the process by which to achieve them. The human user inputs their goal and/or directive, which can implicitly or explicitly set parameters on the process and results. Explainability and an effective user interface (UI) are therefore critical in the design, implementation, and evaluation of human performance assessment AI systems to prevent "garbage-in, garbage-out" scenarios that result in suboptimal and/or discriminatory decision-making that could have been more competently executed by a human alone (Weyerer & Langer, 2019).

In essence, user-centered AI goes beyond mere convenience; it becomes a catalyst for human growth and development. It empowers individuals to perform at their best, offers personalized learning and training experiences, and provides objective assessments while upholding principles of fairness and equity. As AI continues to integrate into various aspects of our lives, its user-centric approach becomes essential for realizing the full potential of AI as a tool for human advancement.

To this end, this chapter specifically examines Explainable Artificial Intelligence (XAI), what it is; its importance to AI-supported human performance training, selection, and assessment; and the issues that need to be overcome to ensure its effective utility in these key application areas. One such issue of major focus is the subject of bias. We herein discuss the nature of bias, its ubiquitous influence on human decision-making, and how said biases may be inadvertently embedded in AI systems (and the dangers of pervasive beliefs that AI is inherently bias-free). Furthermore, we provide concrete examples of how different biases have already infiltrated AI-informed decision support systems and the ensuing societal consequences. We also profess the critical and immediate need for the generation and implementation of standardized ethical guidelines for the design and use of XAI. We provide extensive descriptions and recommendations with regard to the current best practices and guides available for the designer and practitioner. Finally, we describe the largest challenges still facing user-centered XAI systems to guide future efforts for their improvement to promote and safeguard the desirable outcomes of ethical effectiveness, efficiency, and safety of XAI-supported human performance training, selection, and assessment systems.

WHAT IS EXPLAINABILITY?

EXPLAINABILITY (XAI)

With the rise in the complexity of AI applications, calls have been made to improve the quality of explanations provided by these systems. In interactions between humans, explanations are critical for communication because they provide reasonings and justifications for actions. To be useful to humans, the explanations provided by AI must be interpretable (Phillips et al., 2020). However, as algorithms do not "reason" in the same way that humans do, gleaning interpretable, meaningful explanations can be challenging (Lipton, 2018). The field of machine learning (ML) explainability seeks to increase the interpretability of AI models. While recent progress has been made in the field of AI explainability, many models are still opaque, meaning that they do not offer any insights into their algorithmic mechanisms (Doran et al., 2017). The system consequently remains a black box, which increases the chances of "garbage in, garbage out" suboptimal performance.

One burgeoning area in AI explainability, XAI, attempts to articulate the problem, converge on concepts, and provide tools and techniques to verify explainability (Ehsan et al., 2021). The EU Ethical Guidelines for Trustworthy Autonomy (2019) refer to explicability as one of the four ethical principles necessary for trustworthy AI, with the other three being (1) respect for human autonomy, (2) prevention of harm, and (3) fairness. Phillips et al. (2020) describe XAI using four principles in the National Institute of Standards and Technology (NIST) guidance:

- The first principle, explanation, is merely indicating that some explanation should be available – it does not speak to the correctness, informativeness, or intelligibility of said explanation.
- The second principle, meaningfulness, refers to whether the intended audience can understand the explanation. This is necessarily based on an understanding of said audience (Hind, 2019).
- Correctness of the explanation is prescribed in the third principle, explanation accuracy. This concept is independent from explanation accuracy, which refers to whether the judgments made by the system are correct – this guideline refers only to the veracity of the explanations the system provides. This principle is similar to meaningfulness in that it is relative to the expertise and needs of the audience receiving the explanation.
- The fourth principle, knowledge limits, identifies where the system is and is not designed or approved to operate, or where the information it provides is likely to be unreliable. This principle is critical for trust as it prevents misleading outputs and supports ethical AI by preventing unjust outcomes by adhering guidelines based on known limitations.

Transparency, which refers to a clear presentation of the inner workings of a system (Lipton, 2018), is another consequential factor in XAI. Transparency is one of the eight general principles set out in the Institute for Electrical and Electronics Engineers (IEEE) Ethically Aligned Design: "The basis of a particular autonomous and intelligent system decision should always be discoverable" (Shahriari & Shahriari, 2017, p. 4). In an article summarizing the Defense Advanced Research Projects Agency (DARPA) XAI project, Gunning and Aha (2019) describe the need for both explainable models and explanatory interfaces to communicate with the user. In contrast to black-box models that do not provide insight into their algorithmic mechanisms, XAI supports the evolution toward more transparent models, sometimes referred to as white- or glass-box models. Vilone and Longo (2021) describe white-box models as self-explainable and interpretable. Moreover, Rai (2020) describes XAI as a method to turn the opaque, black-box models into glass-box models that are transparent for inspection by humans.

When considering transparency, it is important to acknowledge that it is not only a characteristic of the system – but a product of the human and machine together that occurs through interaction. For transparency to be achieved, the information must be understood by the user – this points to a complex interplay between transparency and interpretability. Transparency is therefore an emergent property that is a product of the human-automation system (Ososky et al., 2014). This concept of communicating in a way the human user can interpret is widely considered a core purpose of XAI (Balasubramaniam et al., 2022). While these guiding principles need to be at the forefront of designers' and practitioners' minds, there are critical issues currently hampering the systems' ability to manifest these key constructs through user-centered design. It is to these imperative concerns that we now turn our focus.

Issues with XAI

While XAI can certainly improve the human-interpretability aspects of AI, it brings its own concerns. Technical complexity, making ML algorithms transparent to adversaries, and trust calibration are all issues associated with XAI. Technical complexity is often described as a barrier to XAI as users often lack the technical expertise to understand the complex coding used to develop the systems; and while XAI seeks to address this issue, many of the methods currently available to operationalize XAI are too technical for most users and stakeholders to fully understand (Bhatt et al., 2020). These authors recommend deeply understanding the intended user audience in question so that explanations and methods can be tailored to their specific needs. Lipton (2018) discusses how despite a lack of a clear definition for interpretability, a growing body of literature has proposed developing algorithms that are interpretable. However, providing a clear definition of interpretability is a challenging proposition considering the diversity of definitions and ideas surrounding the concept.

Other issues associated with XAI concepts involve the proposition to use XAI as a means of increasing trust, which is not necessarily a desirable outcome. Importantly, human trust in ML applications should not be envisioned as a simple dichotomy where the user does or does not trust the system. Whether or not a human user *should* trust a system depends not only on how the system behaves, but also how the system should behave and how the system communicates with the user. That is to say, the goal is not necessarily for the human user to trust the ML application as much as possible, but that the human user trusts the ML application appropriately in accordance with its capabilities – a concept known as trust calibration (Lee & See, 2004). Moreover, presenting more information does not necessarily lead to more trust. For instance, Cheng et al. (2019) provided interactive explanations that aided users' understanding of an ML algorithm, but it did not lead to an increase in trust. This result may be because explainability is only one factor influencing trust in AI, and other factors, such as resiliency, reliability, bias, and accountability (Philips et. al., 2020), may weigh more heavily on the users' trust perceptions. Poor outcomes may be also due to properly calibrated trust (e.g., the system should not be trusted and as the user understands the system, they appropriately place a lower amount of trust in it).

Risks also arise from the desire for transparency. One risk of implementing transparent AI systems is the likelihood of successful adversarial attacks. Making the inner workings of a model more accessible leaves that model more vulnerable to bad actors, who may leverage the explanation to compromise the system (Kuppa & Le-Khac, 2021). Similarly, companies publishing transparent models risk losing their competitive advantage by sharing potentially privileged information or information that can be replicated (Burrell, 2016).

These concerns are all central to AI's functioning, implementation, and security. However, another critical consideration that is fundamental to the operation of these systems is bias. Proponents proselytize that one of the major reasons to implement AI at all is because it will eradicate the bias that so often affects decision-making in humans. The assumption is that you cannot have human biases in a non-human decision-maker. We now discuss the nature of bias in both humans and AI, and confront the extent to which this claim of "bias-free" AI decision-making is true.

BIAS

What Is Bias?

Bias refers to a "systematic difference in treatment of certain objects, people, or groups in comparison to others" (International Organization for Standardization, 2021, p. 1) This concept is relevant to both human behavior and AI development. Biases in human behavior impact the data used in ML models, which can also be impacted by statistical biases.

When discussing bias in humans, we are generally referring to unfair or even prejudicial thinking. However, the domain of cognitive biases examines the cognitive processes involved with human biases more deeply. This term, coined by Tversky and Kahneman in the 1970s, referred to systematic, but potentially flawed behavioral patterns (Wilke & Mata, 2012). Rather than using objective input, humans often rely on their own individual construction of reality to guide their responses. Human biases may be recognized (explicit) or unrecognized (implicit). While explicit biases may be controlled with effort, implicit biases are more complicated because they are not consciously recognized. As these implicit biases are not consciously recognized, they are challenging to measure or quantify. While one test, the Implicit Associations Test (IAT) claims to use response times to stimuli as a quantified measure of implicit attitudes (Greenwald et al., 1998), some scientists feel there is insufficient evidence for their claims (Schimmack, 2021), potentially complicating the construct of implicit bias.

Tversky and Kahneman (1974) described the gap between rational choices and observed human judgments as cognitive bias. Heuristics, which are mental shortcuts based on previous experience that support fast decision-making, play an important role in human judgments. However, these heuristics can and often do lead to errors.

While many individuals assume ML models are objective, data for those models come from the real world and are impacted by human biases. Google's Crash Course in ML describes the following biases that impact ML algorithms:

- Reporting bias: Frequencies represented in data do not reflect real-world frequencies.
- Automation bias: Tendency to favor results from automated systems over result from non-automated systems.
- Selection bias: Sample not reflective of the real-world distribution.
- Group attribution bias: Tendency to generalize observations made about individuals to an entire group.
- Implicit bias: Assumptions made based on personal experience rather than real-world information.

While this is a relatively short sample of the long list of known human biases, it demonstrates the potential impact of those biases on ML.

Developers/Users May Falsely Believe AI Gets Rid of Bias

Developers and users may believe that the use of AI in assessments helps to eliminate bias. However, Manyika et al. (2019) provide several examples of how systems developed to remove bias in assessment failed to do so. They provide several cases in which criminal justice algorithms, hiring algorithms, and facial recognition technologies – that were each designed to unbiasedly assist in assessment and decision-making – resulted instead in hurting and discriminating against already marginalized groups. These technologies were developed with the intent that they would remove the human bias by having an algorithm make an impartial decision; however, each system suffered from a dearth of training data to properly represent all of the populations it would be requested to judge.

Input Matters, Garbage In, Garbage Out, and System Parameters

A lack of training data is not the only issue. Specifically, we can see that training data itself input into the AI matters (Jelly, 2023). Weyerer and Langer (2019) point out that AI developed leveraging bad inputs causes those negative outcomes to propagate throughout the AI lifecycle. This practice can cause perpetual harm as the resulting assessments are often fed back into this cycle. Furthermore, humans are often charged with placing parameters on these AI systems. People can be quite unaware of their lack of knowledge in an area. Though developers may behave altruistically while creating an AI for the assessment of humans, they may be missing one or more large pieces of information needed to build this system without bias. Maniyka et al. (2019) pointed out one such example, the Amazon hiring algorithm was halted when it was discovered that it preferred applicants that used specific verbs like "executed." It is clear to most individuals that virtually no one would expect such an innocuous word – used to describe completing a task – or other similar style words would cause bias, but that was the exact result.

ETHICAL STANDARDS MUST BE BUILT

Guardrails Are Needed for Designers and Policymakers

There are several mitigation opportunities for the disastrous results indicated in the example above that can be considered throughout not only the design process but in policy making as well. We recommend that policy makers and designers start to develop their guardrails beginning with lowest level of their automation and continue throughout the highest level (Parasuraman et al., 2000). However, we also recognize that there are challenges to overcome due to the unique nature of AI technology.

Input These Guardrails throughout the Process and into the Stages of Automation

The development of AI is unlike the development of any previous new technology, particularly regarding the question of ethics. There are a few reasons for this, and those reasons combined are resulting in the exploration of uncharted legislative territory, from both a development and usage standpoint, across a myriad of fields and disciplines.

One of the stark differences in the advancement of AI is in the inability for governmental agencies to apply limitations or moratoria on the direction or rate of progress regarding the technology. To date, government agencies have been unable to properly define what specifically constitutes AI, what forms of the technology will be covered under what not-yet-written legal statutes, and who would be covered under those legal protections (Hacker et al., 2023). In other areas of technological innovation when the question of the ethics of the technology is a cause for concern, it is possible to halt or delay the progress of development until either the experts in the field, or unfortunately more commonly, political authorities have the opportunity to weigh in. This trend has been the case in the sphere of stem cell research starting in the 1970s, with various countries around the globe halting research to have only some countries begin to allow the research to take place with numerous restrictions in place (Hughes, 2021). This type of pause is meant to ensure careful, effective, and ethical progress toward these desirable scientific and societal goals. However, such a moratorium is not feasible in the realm of AI as all the necessary equipment and knowledge to successfully engage in the exploration of AI is widely and readily available to the general public. There is no need for advanced education in a specific domain area; in fact, as the demand for a technologically skilled workforce increases, a significant number of programmers are self-taught, learning through various online resources (Shen, 2020). Additionally, there is no need for expensive laboratory set-ups or specialized equipment as all that is required is access to computers and the software to run them. So, with the inability to completely and effectively regulate the advancement of AI, and which individuals or entities are involved in the processes, how does the structure of a system of ethics even begin to be applied? Who would be the regulatory body to apply that system of ethics and from what culture or society should that regulatory body be informed? It quickly becomes clear that there are no clear-cut answers to these questions and, at this point in time, no obvious path to a solution.

On top of all the issues in the ethical development of AI, it then becomes a question of how the usage of AI will, across the numerous disciplines in which it is applicable and potentially useful, be moral or fair. There are several areas where the use of AI is already questionable (as we discussed at length in an earlier section), and that number is steadily growing. In the field of art, for example, AI has been utilized to expand and explore the artistic vision in a variety of media allowing for increased creativity and expression, but the concern remains, what happens when the AI learns to approximate human inventiveness just enough that the human becomes irrelevant to the monetary exchange for paintings, sculptures, or music (Roose, 2022)? Will humanity, or the output of AI productivity, get to the point where the ability to distinguish the difference between the two is lost (Demmer et al., 2023)? Are artists who put their work out into the world unintentionally training the AI that will replace them? Another form of protected intellectual property at potential risk is the creative forces at work in advertisement. What will be the impact of the use of AI on graphic designers, people who are an integral component of the advertisement world where the profit margin is of utmost importance (Engawi et al., 2021)? Just recently the protections for content creation were won as part of a larger deal that screenwriters in Hollywood negotiated with studios in which AI use must be disclosed and AI cannot be credited with manuscript creation (Coyle, 2023). The rise of generative AI further complicates the issues of creative license.

It is not just in the worlds of art and entertainment that the use of AI has the potential to be of great benefit, as well as significant harm. In the practice of law, work is still being done to properly define AI and describe how, where, and when that forthcoming legal definition will be applied (Schuett, 2019). Additionally, there is an effort to utilize AI to mitigate discrimination, but without a clear understanding of how AI can appropriately be applied to the law, the ethics of this issue are becoming increasingly murky (Miller, 2020).

In law enforcement, the ethical considerations become even more consequential for society as AI is used ostensibly to reduce racial and socioeconomic biases. However, the AI system used must be programmed, it must learn and be trained to look for patterns to identify potential crimes being perpetrated, but who is doing that programming other than flawed and racially biased humans (Berk, 2021)?

Another serious area of concern is the use of AI in academia. As students turn to available forms of AI, such as ChatGPT, to complete homework assignments or take tests, the risk to academic integrity increases. Are the institutions of higher education conferring degrees on individuals that are under educated and under skilled to enter the job market in their respective fields, creating a deficient work force (Eke, 2023)? There is sufficient cause for concern across the many disciplines in which AI is currently being tested or utilized but that does not mean that AI has no place in the workplace or in education. However, it does require that our global society works cooperatively to develop and place a set of guidelines designed to hold the use of AI to an ethical standard. We now discuss the best nascent efforts to establish and implement guidelines for the ethical design and use of AI systems.

STATE OF PRESENT BEST PRACTICES

Guides for AI Ethics

There is currently a call in the technology industry to adhere to guidelines related to ethical AI development. President Joe Biden stated "We must be clear-eyed and vigilant about the threats emerging from emerging technologies that can pose — don't have to but can pose — to our democracy and our values," during an announcement that seven leading AI companies in the United States had agreed to implement voluntary safeguards on the development of AI technology, including "security testing, in part by independent experts; research on bias and privacy concerns; information sharing about risks with governments and other organizations; development of tools to fight societal challenges like climate change; and transparency measures to identify A.I.-generated material" (Shear et al., 2023, para. 11).

While this agreement clearly highlights the importance of ethics in AI development and the commitment of these companies to consider ethics during this process, the agreements here are somewhat vague. Lack of specificity is one of the issues identified and stressed in recent criticisms of AI guidelines. As Wei and Zhou (2022) observe, though "governments and corporations have curated multiple AI ethics guidelines to curb unethical behavior of AI, the effect has been limited, probably due to the vagueness of the guidelines" (p. 1). In a scathing description of the plethora of AI guidelines released in recent years, Munn (2023) describe meaningless (e.g., vague, abstract, and incoherent), isolated (e.g., lacking social and cultural context), and toothless (e.g., lacking enforcement or consequences) principles that divert resources from more effective outcomes. AlgorithmWatch (2020) also questioned whether unenforceable guidelines were indeed better than having no guidelines at all after compiling over 160 guidelines into a searchable inventory. They reported finding vague formulations and a lack of enforcement mechanisms, as well as a limited perspective emanating mainly from Europe and the United States. Development of guidelines by self-interested parties is also a concern mentioned, indicating that these guidelines should be "more than a PR tool for companies and governments" (para. 6). In the same vein, McMillan and Brown (2019) warn of "ethics washing," which criticizes guidelines development for potentially "diluting our rights in practice, and downplaying the role of our own self-interest" (p. 1), often developed for the goal of avoiding regulation and manipulating public opinion. Inadequate or faulty guidelines can also lead to "ethical debt," wherein AI systems are developed under the presumption that the AI solution itself is ethical. As the development focus is on efficiency and there are often no viable means to address, or even realize the ethical issues, harmful consequences manifest and then must be mitigated (Dorton et al., 2023).

To begin to address some of these issues, Wei and Zhou (2022) evaluated real-world complaints from the AI Incident Database, a catalogue of repetitive AI failures. They created a taxonomy using 150 incidents occurring from 2010 to 2021 that they classify into eight categories: inappropriate use (bad performance), racial discrimination, physical safety, unfair algorithm (evaluation), gender discrimination, privacy, unethical use (illegal use), and mental health. The number one issue they identified was transparency, followed by justice and fairness, and finally non-maleficence, relating to security, safety, harm, and protection. While these three categories align well with the many of the published guidelines on Ethical AI, these authors suggest their taxonomy moves the issues from the vague theoretical realm into practical, concrete terms by describing the consequences of non-compliance with guidelines.

While imperfect, ethical guidelines for AI development may still be useful resources for companies, developers, and anyone wanting to understand the ethical implications of AI development. As mentioned above, there are many guidelines and describing them all is beyond the scope of this chapter, but some of the more prominent guidance is described below, including:

- European Commission (EC) Ethics Guidelines for Trustworthy AI
- Google Responsible AI Practices
- Institute of Electrical and Electronics Engineers (IEEE) AI Ethics and Governance Standards
- International Business Machines (IBM) AI Ethics Guide
- United Nations Educational, Scientific and Cultural Organization (UNESCO) Ethics of AI
- United States Department of Defense (US DoD) Ethical Principles for AI
- United States Intelligence Community (USIC) Principles of AI Ethics and Framework for the Intelligence Community

Table 5.1 summarizes these guides in terms of their purpose, the concerns they address, and what they require, recommend, or provide.

While these standards vary widely in scope, purpose, and detail, some commonalities can be observed. Nearly all of the guidance listed here mentions transparency, interpretability, or explainability. Many also describe the importance of protecting humans, the role of the human, or the implementation of a human-centered design approach, highlighting the importance of AI as a tool for human use. To ensure said tool functions as intended (and as effectively as possible), we now turn to the largest challenges currently affecting the design, use, and implementation of XAI systems intended to improve human training, selection, and assessment.

BIGGEST CHALLENGES TO OVERCOME PERTAINING TO HUMAN PERFORMANCE, ASSESSMENT, TRAINING, AND SELECTION

Though the temptation to leverage AI in human assessment is strong, the moral and ethical implications of doing so without consideration of the challenges this introduces are grave. Specifically, there are multiple challenges associated with the proper development of user-centered AI technologies in the assessment of others (e.g., to select the best performer), as well as in the assessment of one's self for personal consumption (e.g., fitness trackers). First, let us distinguish between the focus

TABLE 5.1
Guides for Ethical AI

Guideline	Source	Purpose(s)	Concern(s) Addressed	Requirements, Recommendations, or Provisions
European Commission (EC) Ethics Guidelines for Trustworthy AI	European Commission, 2019	This guide serves to guide development, deployment, and use of AI	Respect for human autonomy, prevention of harm, fairness and explicability	• Human agency and oversight • Technical robustness and safety • Privacy and data governance • Transparency • Diversity, non-discrimination and fairness • Environmental and societal well-being • Accountability
Google Responsible AI Practices	Google, 2023	Provides general recommendations for AI design, as well as specific guidance on fairness, interpretability, privacy, and safety	General	• Use a human-centered design approach • Identify multiple metrics to assess training and monitoring • When possible, directly examine your raw data • Understand the limitations of your dataset and model • Test, Test, Test • Continue to monitor and update the system after deployment
Google Responsible AI Practices	Google, 2023	Provides general recommendations for AI design, as well as specific guidance on fairness, interpretability, privacy, and safety	Interpretability	• Plan out your options to pursue interpretability • Treat interpretability as a core part of the user experience • Design the model to be interpretable • Choose metrics to reflect the end-goal and the end-task • Understand the trained model • Communicate explanations to model users • Test, Test, Test
Google Responsible AI Practices	Google, 2023	Provides general recommendations for AI design, as well as specific guidance on fairness, interpretability, privacy, and safety	Privacy	• Collect and handle data responsibly • Leverage on-device processing where appropriate • Appropriately safeguard the privacy of ML models

Google Responsible AI Practices	Google, 2023	Provides general recommendations for AI design, as well as specific guidance on fairness, interpretability, privacy, and safety	Security	• Identify potential threats to the system • Develop an approach to combat threats • Keep learning to stay ahead of the curve
IEEE Recommended Practice for Assessing the Impact of Autonomous and Intelligent Systems on Human Well-Being 7010-2020	IEEE, 2020	Envisioned to help developers incorporate human-centric design principles and raise awareness of ethical issues	Assessing the Impact of Autonomous and Intelligent Systems on Human Well-Being	• Product development guidance • Identification of areas for improvement • Risk management • Performance assessments • Support for identification of intended and unintended users • Impacts on human well-being
IEEE Standard for Transparency of Autonomous Systems 7001-2021	IEEE, 2021a	Focuses on the objective assessment of transparency, to operationalize the necessity of users understanding how and why AI makes decisions	Transparency	• How to approach transparency • Requirements by stakeholder and level
IEEE Standard Model Process for Addressing Ethical Concerns during System Design 7000-2021	IEEE, 2021b	Provides processes that support consideration of ethical values during concept exploration and development through stakeholder engagement	Ethical values during concept exploration and development through stakeholder engagement	• Identifies key roles, supports ConOps development, feedback elicitation and prioritization, ethical requirements definition, and describes ethical risk-based design processes, and transparency management processes.
International Business Machines (IBM) AI Ethics Guide	IBM, 2023	AI guide to ethics for data scientists and researchers to support ethical AI development to benefit the greater society	Support ethical AI development to benefit the greater society, including governance and explainability	Leveraging the principles of the Belmont report (respect for persons, beneficence, and justice) to guide AI development, describes contemporary concerns regarding AI, and describes how to establish ethical guidelines; provides a structure for establishing AI ethics that includes governance and explainability

(Continued)

TABLE 5.1 *(Continued)*
Guides for Ethical AI

Guideline	Source	Purpose(s)	Concern(s) Addressed	Requirements, Recommendations, or Provisions
United Nations Educational, Scientific and Cultural Organization (UNESCO) Ethics of AI	UNESCO, 2021	Focused on human rights; highlights the importance of human oversight of AI through the advancement of principles such as fairness and transparency	Supports policy development	Four core values supporting AI dev for the good of humanity: 1. Human rights and human dignity: Respect, protection and promotion of human rights and fundamental freedoms and human dignity 2. Living in peaceful just, and interconnected societies 3. Ensuring diversity and inclusiveness 4. Environment and ecosystem flourishing Principles: • Proportionality and Do No Harm • Safety and security • Fairness and non-discrimination • Sustainability • Right to Privacy, and Data Protection • Human oversight and determination • Transparency and explainability • Responsibility and accountability • Awareness and literacy • Multi-stakeholder and adaptive governance and collaboration
United States Department of Defense (US DoD) Ethical Principles for AI	U.S. Department of Defense, 2021	concise set of ethical principles for the use of AI in accordance with America's commitment to responsibility and lawful behavior	Supports innovation and advance trustworthy AI while upholding DoD ethical standards	Dictate AI should be: • Responsible • Traceable • Reliable • Governable

United States Intelligence Community (USIC) Principles of AI Ethics and Framework for the Intelligence Community	Office of the Director of National Intelligence, 2021a,b	Governs design, development and use of AI in the USIC through ethical principles	Framework for what AI should do that governs design, development and use of AI in the USIC through ethical principles	• Respect the Law and Act with Integrity • Transparent and Accountable • Objective and Equitable • Human-Centered Development and Use • Secure and Resilient • Informed by Science and Technology
United States Intelligence Community (USIC) Principles of AI Ethics and Framework for the Intelligence Community	Office of the Director of National Intelligence, 2021a,b	Governs design, development and use of AI in the USIC through ethical principles	Guidance for individuals who design, develop, review, deploy, and use AI are sufficiently trained to address the following issues regarding how AI should be used	• Be used when it is an appropriate means to achieve a defined purpose after evaluating the potential risks; • Be used in a manner consistent with respect for individual rights and liberties of affected individuals, and use data obtained lawfully and consistent with legal obligations and policy requirements; • Incorporate human judgment and accountability at appropriate stages to address risks across the lifecycle of the AI and inform decisions appropriately; • Identify, account for, and mitigate potential undesired bias, to the greatest extent practicable without undermining its efficacy and utility; • Be tested at a level commensurate with foreseeable risks associated with the use of the AI; • Maintain accountability for iterations, versions, and changes made to the model; • Document and communicate the purpose, limitation(s), and design outcomes; • Use explainable and understandable methods, to the extent practicable, so that users, overseers, and the public, as appropriate, understand how and why the AI generated its outputs; • Be periodically reviewed to ensure the AI continues to further its purpose and identify issues for resolution; and, • Identify who will be accountable for the AI and its effects at each stage and across its lifecycle, including responsibility for maintaining records created.

(Continued)

TABLE 5.1 *(Continued)*
Guides for Ethical AI

Guideline	Source	Purpose(s)	Concern(s) Addressed	Requirements, Recommendations, or Provisions
United States Intelligence Community (USIC) Principles of AI Ethics and Framework for the Intelligence Community	Office of the Director of National Intelligence, 2021a,b	Governs design, development and use of AI in the USIC through ethical principles	Issues individuals should be able to address	• Understanding Goals and Risks • Legal Obligations and Policy Considerations Governing the AI and the Data. • Human Judgment and Accountability • Mitigating Undesired Bias and Ensuring Objectivity. • Testing Your AI • Accounting for Builds, Versions, and Evolutions of an AI • Documentation of Purpose, Parameters, Limitations, and Design Outcomes • Transparency: Explainability and Interpretability • Periodic Review • Stewardship and Accountability: Training Data, Algorithms, Models, Outputs of the Models, Documentation

on user-centered and human-centered AI (HCAI). We define user-centered AI as those considerations focused on the consumer or user of the AI technology, whereas HCAI can be thought of as more holistic to consider the full partnership of humans and AI to foster human agency (Ozmen-Garibay et al., 2023).

Ozmen-Garibay et al. (2023) completed an extensive literature review to highlight and categorize several of the major issues associated with HCAI. They emphasize challenges in the six areas of human well-being, responsible design practices, privacy considerations, human-centered design principles, oversight, and respect for human cognition as the largest hurdles to human-centered AI development. Though all of these HCAI challenges apply to the use of AI in the assessment of human training, performance, and selection in some way, we will focus on how they apply to our topic of user-centric issues. We take this approach from two perspectives – the assessment of others and the assessment of the self. Further, we argue that from an ethical and moral standpoint, human well-being should serve as the ultimate goal of deploying AI in performance assessment, thus we describe how five of the categories identified by Ozmen-Garibay et al. can serve to enhance or undermine well-being.

Human well-being, or psychological well-being (PWB), is described as personal perceptions of one's own levels of environmental mastery, autonomy, self-acceptance, personal-growth, life-purpose, and positive relationships with others (Ryff & Keyes, 1995). Closely related to PWB is a theory of motivation, Self-Determination Theory (SDT). Self-Determination Theory posits that human motivation is determined by the fulfillment of three basic psychological needs, including competence, autonomy, and relatedness (Ryan, 2009; Ryan & Deci, 2000). Each of the facets of PWB and SDT can be beneficially or detrimentally impacted by interactions with one's environment. Let us now explore these theories in conjunction with using AI in the assessment of human performance, training and selection from the perspective of the user assessing the self and others. To illustrate the potential challenges, we complete this exploration through the use of two examples. In assessing one's self, we use the example of an AI tool to evaluate physical training results as impacting one's own health. In assessing others, we examine the use of an AI tool to assess physical training results of others for the selection of spots on a sports team.

We can glean insight into how PWB and SDT can be negatively impacted by the use of these technologies by reflecting on a few of the major challenges mentioned by Ozmen-Garibay et al. (2023). Responsible design practices should involve efforts to reduce bias in the training data that contribute to the development of any health technology (Challen et al., 2019), including that of the AI fitness tracking technology. If in development, bias was not accounted for, or could not be predicted, the technology itself may act on an assumption that does not hold true. For example, most fitness tracking devices request a person to input whether they are female, male, other, or prefer not to say. These feature selections have a direct effect on the algorithm used to calculate a person's caloric needs and calories burned based on the activity the tracker and its accompanying application leveraged. However, the trackers are subject to the bias of the data that went into building them, such as the numbers of females, males, others, and those that preferred not to say, in addition to the individual differences each of these persons brought to the data. The resulting AI may not adapt to the nuances of the active user well enough to provide them with accurate caloric advice, thus leading the user to over-eat, under-eat, and gain or lose weight unintentionally resulting in poor PWB (possible distress from unintentional weight changes) and SDT (loss of motivation to try to change one's weight to a healthier level) outcomes. This case study is only one small example of how lack of responsible design, despite good intention, can cause bias and have a negative impact on PWB and SDT.

As stated above, fitness trackers and their applications require personal data (Yang, 2021), there are consequently privacy considerations that come into play that can impact PWB and SDT. Above and beyond the individual feature data these technologies request, many also require permissions and/or access to other data on the device the application runs on. For example, the device or application may access phone calls, messaging, location, and/or a host of other information. Of course, many of these serve to help the tracker and application to run properly – it would be difficult to tell you how far you ran outdoors during a workout without accessing your GPS location – however, the tracker may also be logging information such as which stores you visit and where you spend your free time. Though these data collection points are circumstantial, user environmental mastery and self-acceptance can be negatively impacted if the user does not realize that these data may be used to target advertisements to them, especially if those ads are of a sensitive nature. This hypothetical example represents just a small way we believe that lack of privacy considerations during AI development can impact PWB and SDT for users leveraging the technology to assess themselves.

Our next hypothetical scenario is in the use of an AI technology assessing the performance of others for selection of a position on a sports team. It makes sense that it can be appealing to leverage such technology in this type of selection as one could argue that it might reduce human bias or favoritism in the process. However, we know that trust in the technology can affect someone's acceptance of the information it provides (Lee & See, 2004). Specifically, if user trust is not properly calibrated to or matching the abilities of a technology, misuse, disuse, and/or abuse of the technology can result. In our example, if the technology is not designed in a human-centered way, and the user has too much trust in the technology's ability, that person may accept what the technology says without question. This outcome could result in the wrong players being selected for the team, negatively impacting the PWB and SDT through unsupported knocks on the competence and self-acceptance of those who were actually the better choices. Furthermore, the person leveraging the technology might become complacent with the technology choices, thus lessening their autonomy and environmental mastery, impacting their PWB and SDT.

Continuing with this example set, designers of such an AI technology need to maintain a respect for the user's cognition in the task of selecting members of their team. Many current developers have been given a goal of creating explainable AI to support user cognition. However, *explainable* does not always translate to *understandable* (Herm et al., 2023). When reviewing data from each of the people trying out for the sports team, the AI needs to produce information that is digestible to the person that will be ingesting it. More specifically, if the technology discusses the type of algorithm it used to achieve its decision or recommendation, it is likely not supporting the average user; it is undermining competence, and negatively impacting

PWB and SDT. It should rather point to clear, easy to understand metrics that distinguish its decisions and recommendations. This example set certainly does not cover the wide range of ways that leveraging AI for performance, training, and selection activities can cause negative impacts as a full example set would be a book in and of itself. However, we invite and challenge our readers to consider these cases as inspiration to further examine how AI can and does impact their daily lives.

THE CURRENT LARGEST SUCCESSES/BENEFITS

We have pointed out just a few of the many ways this type of technology can harm those involved; however, there are strong benefits that in leveraging AI for assessment and selection. AI has the ability to accept a very large number of inputs from a series of complex data points simultaneously and can provide a real-time or near-real-time assessment of these inputs that would otherwise take humans a significant amount of time to process. Specifically, Alowais et al. (2023) believe that AI has the ability to revolutionize healthcare diagnostics and treatment plans. These researchers argue that as long as there is a proper consideration for the potential bias and data privacy issues that will arise, the ability of AI to sift through data to assess health conditions and provide a series of potential diagnoses, as well as select an optimal series of treatment plans for these diagnoses can and will speed along medical intervention. However, we again caution those wishing to use these types of technologies – these are not and should not be relied upon as a replacement or a crutch for proper medical education and practice – rather, they must be used a series of tools to augment the knowledge and creativity of excellent practitioners. We advise that this caution extend to all instances of AI leveraged in the training, selection, and/or assessment of humans.

MITRE disclaimer: Approved for public release. Distribution unlimited PR_24-02288-1

REFERENCES

AlgorithmWatch. (2020). In the realm of paper tigers – exploring the failings of AI ethics guidelines. https://algorithmwatch.org/en/ai-ethics-guidelines-inventory-upgrade-2020/

Alowais, S. A., Alghamdi, S. S., Alsuhebany, N., Alqahtani, T., Alshaya, A. I., Almohareb, S. N., …, & Albekairy, A. M. (2023). Revolutionizing healthcare: The role of artificial intelligence in clinical practice. *BMC Medical Education*, *23*(1), 689.

Balasubramaniam, N., Kauppinen, M., Hiekkanen, K., & Kujala, S. (2022, March). Transparency and Explainability of AI Systems: Ethical Guidelines in Practice. In *International Working Conference on Requirements Engineering: Foundation for Software Quality* (pp. 3–18). Cham: Springer International Publishing.

Berk, R. A. (2021). Artificial intelligence, predictive policing, and risk assessment for law enforcement. *Annual Review of Criminology*, *4*(1), 209–237. https://doi.org/10.1146/annurev-criminol-051520-012342

Bhatt, U., Xiang, A., Sharma, S., Weller, A., Taly, A., Jia, Y., …, & Eckersley, P. (2020, January). Explainable machine learning in deployment. In *Proceedings of the 2020 conference on fairness, accountability, and transparency* (pp. 648–657).

Budhwar, P., Chowdhury, S., Wood, G., Aguinis, H., Bamber, G. J., Beltran, J. R., …, & Varma, A. (2023). Human resource management in the age of generative artificial intelligence: Perspectives and research directions on ChatGPT. *Human Resource Management Journal*, *33*(3), 606–659.

Burrell, J. (2016). How the machine 'thinks': Understanding opacity in machine learning algorithms. *Big Data & Society*, *3*(1), 2053951715622512.

Challen, R., Denny, J., Pitt, M., Gompels, L., Edwards, T., & Tsaneva-Atanasova, K. (2019). Artificial intelligence, bias and clinical safety. *BMJ Quality & Safety*, *28*(3), 231–237.

Chen, B., Wu, Z., & Zhao, R. (2023). From fiction to fact: The growing role of generative AI in business and finance. *Journal of Chinese Economic and Business Studies*, *21*(4), 471–496.

Cheng, H. F., Wang, R., Zhang, Z., O'connell, F., Gray, T., Harper, F. M., & Zhu, H. (2019, May). Explaining decision-making algorithms through UI: Strategies to help non-expert stakeholders. In *Proceedings of the 2019 chi conference on human factors in computing systems* (pp. 1–12).

Coyle, J. (2023, September 27). *In Hollywood writers' battle against AI, humans win (for now)*. ABC News. https://abcnews.go.com/Business/wireStory/hollywood-writers-battle-ai-humans-win-now-103543408

Demmer, T. R., Kühnapfel, C., Fingerhut, J., & Pelowski, M. (2023). Does an emotional connection to art really require a human artist? Emotion and intentionality responses to ai- versus human-created art and impact on aesthetic experience. *Computers in Human Behavior*, *148*, 107875. https://doi.org/10.1016/j.chb.2023.107875

Doran, D., Schulz, S., & Besold, T. R. (2017). What does explainable AI really mean? A new conceptualization of perspectives. *arXiv preprint arXiv:1710.00794*.

Dorton, S., Ministero, L. M., Alaybek, B., & Bryant, D. J. (2023). Foresight through naturalistic tools for ethical AI. *Frontiers in Artificial Intelligence*, *6*, 1143907.

Ehsan, U., Liao, Q. V., Muller, M., Riedl, M. O., & Weisz, J. D. (2021, May). Expanding explainability: Towards social transparency in ai systems. In *Proceedings of the 2021 CHI Conference on Human Factors in Computing Systems* (pp. 1–19).

Eke, D. O. (2023). Chatgpt and the rise of Generative AI: Threat to academic integrity? *Journal of Responsible Technology*, *13*, 100060. https://doi.org/10.1016/j.jrt.2023.100060

Engawi, D., Gere, C., & Richards, D. (2021, December). *The impact of artificial intelligence on graphic design: Exploring the challenges and possibilities of AI-driven autonomous branding*. SpringerLink. https://link.springer.com/chapter/10.1007/978-981-19-4472-7_238

European Commission, 2019. Ethics guidelines for trustworthy AI. European Commission: High-Level Expert Group on Artificial Intelligence.

Google. (2023) Advancing AI for everyone. https://ai.google

Google. (2023). Fairness: Types of bias https://developers.google.com/machine-learning/crash-course/fairness/types-of-bias

Greenwald, A. G., McGhee, D. E., & Schwartz, J. L. (1998). Measuring individual differences in implicit cognition: The implicit association test. *Journal of Personality and Social Psychology*, *74*(6), 1464.

Gunning, D., & Aha, D. W. (2019). DARPA's explainable artificial intelligence program. *AI Mag*, *40*(2), 44.

Hacker, P., Engel, A., & Mauer, M. (2023). Regulating CHATGPT and other large generative AI models. *2023 ACM Conference on Fairness, Accountability, and Transparency*. https://doi.org/10.1145/3593013.3594067

Herm, L. V., Heinrich, K., Wanner, J., & Janiesch, C. (2023). Stop ordering machine learning algorithms by their explainability! A user-centered investigation of performance and explainability. *International Journal of Information Management*, *69*, 102538.

Hind, M. (2019). Explaining explainable AI. *XRDS: Crossroads, The ACM Magazine for Students*, *25*(3), 16–19.

Hughes, V. (2021). National determinants of human embryonic stem cell research policy in select countries. *American Journal of Public Health Research*, *10*(1), 11–21. https://doi.org/10.12691/ajphr-10-1-3

IBM (2023) What are AI ethics? https://www.ibm.com/topics/ai-ethics

IEEE Standards Association. (2020). IEEE Recommended Practice for Assessing the Impact of Autonomous and Intelligent Systems on Human Well-Being: IEEE Standard 7010-2020. IEEE.

IEEE Standards Association. (2021a).IEEE Standard for Transparency of Autonomous Systems: IEEE Standard 7001-2021. IEEE.

IEEE Standards Association. (2021b). IEEE Standard Model Process for Addressing Ethical Concerns during System Design: IEEE Standard 7000-2021. IEEE.

International Organization for Standardization. (2021). *Information technology — Artificial intelligence (AI) — Bias in AI systems and AI aided decision making* (ISO Standard No. 24027:2018). https://www.iso.org/obp/ui/en/#iso:std:iso-iec:tr:24027:ed-1:v1:en

Jelly, S., (2023). Garbage in, garbage out: The role of data management in effective AI. *Forbes*. Retrieved from https://www.forbes.com/sites/forbesbusinesscouncil/2023/11/16/garbage-in-garbage-out-the-role-of-data-management-in-effective-ai/

Kulkarni, S., Seneviratne, N., Baig, M. S., & Khan, A. H. A. (2020). Artificial intelligence in medicine: Where are we now? *Academic Radiology*, *27*(1), 62–70.

Kuppa, A., & Le-Khac, N. A. (2021). Adversarial xai methods in cybersecurity. *IEEE Transactions on Information Forensics and Security*, *16*, 4924–4938.

Lee, J. D., & See, K. A. (2004). Trust in automation: Designing for appropriate reliance. *Human Factors*, *46*(1), 50–80.

Lipton, Z. C. (2018). The mythos of model interpretability: In machine learning, the concept of interpretability is both important and slippery. *Queue*, *16*(3), 31–57.

Manyika, J., Silberg, J., & Presten, B. (2019, October 25). What do we do about the biases in AI?. *Harvard Business Review*. Retrieved from https://hbr.org/2019/10/what-do-we-do-about-the-biases-in-ai

McMillan, D., & Brown, B. (2019, November). Against ethical AI. In *Proceedings of the Halfway to the Future Symposium 2019* (pp. 1–3).

Miller, K. (2020). A matter of perspective. *Legal Regulations, Implications, and Issues Surrounding Digital Data*, 182–202. https://doi.org/10.4018/978-1-7998-3130-3.ch010

Munn, L. (2023). The uselessness of AI ethics. *AI and Ethics*, 3(3), 869–877.

Nemorin, S., Vlachidis, A., Ayerakwa, H. M., & Andriotis, P. (2023). AI hyped? A horizon scan of discourse on artificial intelligence in education (AIED) and development. *Learning, Media and Technology*, *48*(1), 38–51.

Office of the Director of National Intelligence. (2021a). The IC Principles of Artificial Intelligence Ethics. https://www.dni.gov/files/ODNI/documents/Principles_of_AI_Ethics_for_the_Intelligence_Community.pdf

Office of the Director of National Intelligence. (2021b). The IC Artificial Intelligence Ethics Framework. https://www.dni.gov/files/ODNI/documents/AI_Ethics_Framework_for_the_Intelligence_Community_10.pdf

Ososky, S., Sanders, T., Jentsch, F., Hancock, P., & Chen, J. Y. (2014, June). Determinants of system transparency and its influence on trust in and reliance on unmanned robotic systems. *Unmanned Systems Technology XVI*, *9084*, 112–123.

Ozmen-Garibay, O., Winslow, B., Andolina, S., Antona, M., Bodenschatz, A., Coursaris, C., …, & Xu, W. (2023). Six human-centered artificial intelligence grand challenges. *International Journal of Human–Computer Interaction*, *39*(3), 391–437.

Parasuraman, R., Sheridan, T. B., & Wickens, C. D. (2000). A model for types and levels of human interaction with automation. *IEEE Transactions on Systems, Man, and Cybernetics-Part A: Systems and Humans*, *30*(3), 286–297.

Phillips, P. J., Hahn, A. C., Fontana, P. C., Broniatowski, D. A., & Przybocki, M. A. (2020). Four principles of explainable artificial intelligence (draft).

Rai, A. (2020). Explainable AI: From black box to glass box. *Journal of the Academy of Marketing Science*, *48*, 137–141.

Reiling, A. D. (2020). Courts and artificial intelligence. *International Journal for Court Administration*, *11*(2). https://doi.org/10.36745/ijca.343

Roose, K. (2022, September 2). *An a.i.-generated picture won an art prize. artists aren't happy.* The New York Times. https://www.nytimes.com/2022/09/02/technology/ai-artificial-intelligence-artists.html

Ryan, R. (2009). Self determination theory and well being. *Social Psychology*, *84*(822), 848.

Ryan, R. M., & Deci, E. L. (2000). Self-determination theory. Basic psychological needs in motivation, development, and wellness.

Ryff, C. D., & Keyes, C. L. M. (1995). The structure of psychological well-being revisited. *Journal of Personality and Social Psychology*, *69*(4), 719.

Sætra, H. S. (2023). Generative AI: Here to stay, but for good? *Technology in Society*, *75*, 102372.

Schimmack, U. (2021). The implicit association test: A method in search of a construct. *Perspectives on Psychological Science*, *16*(2), 396–414.

Schuett, J. (2019). A legal definition of AI. *SSRN Electronic Journal*. https://doi.org/10.2139/ssrn.3453632

Searle, J. R. (1980). Minds, brains, and programs. *Behavioral and Brain Sciences*, *3*(3), 417–424.

Shahriari, K., & Shahriari, M. (2017, July). IEEE standard review—Ethically aligned design: A vision for prioritizing human wellbeing with artificial intelligence and autonomous systems. In *2017 IEEE Canada International Humanitarian Technology Conference (IHTC)* (pp. 197–201). IEEE.

Shear, M. D., Kang, C., & Sanger, D. E. (2023). https://www.nytimes.com/2023/07/21/us/politics/ai-regulation-biden.html

Shen, R. (2020). Interactive computer tutors as a programming educator: Improving learners' experiences. *2020 IEEE Symposium on Visual Languages and Human-Centric Computing (VL/HCC)*. https://doi.org/10.1109/vl/hcc50065.2020.9127281

Stein, A. L. (2020). Artificial intelligence and climate change. *Yale Journal on Regulation*, *37*, 890–939.

Tversky, A., & Kahneman, D. (1974). Judgment under uncertainty: Heuristics and biases: biases in judgments reveal some heuristics of thinking under uncertainty. *Science*, *185*(4157), 1124–1131.

UNESCO. (2021). Recommendation on the ethics of artificial intelligence. United Nations Educational, Scientific and Cultural Organization.

U.S. Department of Defense. (2020, February). DOD Adopts Ethical Principles for Artificial Intelligence. https://www.defense.gov/News/Releases/release/article/2091996/dod-adopts-ethical-principles-for-artificial-intelligence/

Vilone, G., & Longo, L. (2021). Classification of explainable artificial intelligence methods through their output formats. *Machine Learning and Knowledge Extraction*, *3*(3), 615–661.

Wei, M., & Zhou, Z. (2022). Ai ethics issues in real world: Evidence from ai incident database. *arXiv preprint arXiv:2206.07635.*

Weyerer, J. C., & Langer, P. F. (2019, June). Garbage in, garbage out: The vicious cycle of ai- based discrimination in the public sector. In *Proceedings of the 20th Annual International Conference on Digital Government Research* (pp. 509–511).

Wilke, A., & Mata, R. (2012). Cognitive Bias. In V. S. Ramachandran (Ed.), *Encyclopedia of Human Behavior*. Cambridge: Academic Press.

Yang, Q. (2021). Toward responsible AI: An overview of federated learning for user-centered privacy-preserving computing. *ACM Transactions on Interactive Intelligent Systems (TiiS)*, *11*(3–4), 1–22.

6 Using Artificial Intelligence to Train Human Intelligence
Theory and Practice in the Design of Adaptive Training Systems

Bradford L. Schroeder, Jason E. Hochreiter, and Wendi L. Van Buskirk

WHAT IS ADAPTIVE TRAINING?

Adaptive training (AT) allows for the emulation of a one-on-one human tutoring experience for a learner (Bloom, 1984; VanLehn, 2011). AT systems are computer systems that allow for the adaptation of a large host of variables that have been deemed relevant for the individual learner's mastery of a topic. These variables include but are not limited to aptitudes (Cronbach, 1957; Park & Lee, 2004), learning preferences or styles (though this has largely been debunked; see Pashler et al., 2009; Nancekivell et al., 2021), instructional interventions such as scaffolding and feedback (Fraulini et al., 2024; Landsberg et al., 2016; Schroeder et al., 2023), and performance (Kelley, 1969; Marraffino et al., 2021). Further, these variables can be used prior to training, to adapt the learning experience during training, or at the end of a training session (Landsberg et al., 2012a).

Before going further to discuss AT systems in detail, we want to clarify terms that are oft confused with AT. These terms include constructs such as adaptive aiding, adaptability, adaptive performance, and adaptive learning. First, adaptive automation involves some form of automation to assist a human on an operational task when they lose processing capacity from increases in workload, fatigue, etc. (Wickens & Hollands, 2000). Specifically, adaptive aiding is a type of adaptive automation that aids the user with a task instead of fully taking over the task. Widely known examples of adaptive aiding in driving are adaptive cruise control and blind spot monitoring. These capabilities do not take over the task of driving but aid the human in driving more safely.

Next, adaptability and adaptive performance are related terms, which focus on how humans can adapt their behavior on the job (Huang et al., 2014). For instance,

DOI: 10.1201/9781032701639-6

Pulakos et al. (2000) define adaptive performance as the proficiency with which an individual alters his or her behavior in response to the demands of a new task, event, unpredictable situation, learning new technology, or environmental constraints (e.g., noise and climates). Adaptability (also sometimes referred to as adaptivity or adaptive learning) is defined as "cognitive, behavioral, and emotional regulation that assists individuals in effectively responding to change, variability, novelty, uncertainty, and transition" (Martin, 2017; p. 696). Therefore, researchers in the domain of adaptability and adaptive performance are investigating the individual difference variable. For instance, researchers are concerned with measurement of the construct (van Dam & Meulders, 2021), understanding the antecedents and consequences in order to predict or improve job performance (Martin, 2017), and/or teaching learners how to be adaptive on the job (Allworth & Hesketh, 1999). This is an entirely different perspective than AT. With adaptivity/adaptive performance, the focus is on *how the human changes* their behavior based on the task environment. In AT, *the system is changing* to meet the needs of the human.

Subsequently, AT is also confused with adaptive learning systems. However, the confusion here is not as problematic as the terms are at least somewhat related. The term adaptive learning systems stems from the artificial intelligence (AI) domain. In this field, adaptive learning systems focus on algorithms or models that learn from a continual influx of data (Zliobaite et al., 2012). Therefore, it is possible that an AT system is also an adaptive learning system by incorporating adaptive learning algorithms or models. However, it is not a requirement for an AT system to be an adaptive learning system. Indeed, most fielded AT systems have rule-based instructional algorithms yet are still quite successful at improving learning gains (Billings, 2012; Johnson et al., 2019; Landsberg et al., 2012b; Van Buskirk et al., 2019).

Finally, the term that is most wrought with confusion is adaptive learning. The term adaptive learning is frequently used synonymously with adaptability/adaptive performance, AT, *and* adaptive learning systems. For this reason, we will refer to instructional systems that adapt to the needs of the learner as AT systems to, hopefully, eliminate the confusion surrounding this construct. Table 6.1 presents a summary of these commonly confused terms.

HOW TO ADAPT TRAINING

At their core, all AT systems must be able to observe the learner's behavior, assess that behavior (i.e., give meaning to that behavior), and provide an instructional response in a way that changes the training for the learner. This is known as the "Observe, Assess, Respond" (OAR) model (Campbell, 2014). In simple terms, following this process generates the algorithms that underpin an AT system.

The complexity of each of these components (i.e., observation, assessment, and response) and their associated algorithms can vary greatly (Campbell, 2014). For example, on the simplistic side, if an elementary school student achieves a 40% score on a math quiz covering fractions (observation), then the system could determine that threshold for passing was not achieved (assessment) and then provide the student with remedial information on what questions they missed (instructional response) before moving them onto the next lesson. On the higher complexity side,

TABLE 6.1
Summary of Terms Commonly Confused with Adaptive Training

Term	Definition	What Adapts?
Adaptability/adaptivity	Changes in thoughts, behaviors, and emotions that help humans respond to new challenges	Human adapts to task
Adaptive aiding	A type of adaptive automation which aids a human with a task instead of fully taking over	System adapts to human
Adaptive automation	Some form of automation that assists a human in completing a task when processing capacity diminishes	System adapts to human
Adaptive learning	Sometimes used to mean adaptability, adaptive performance, adaptive training, *and* adaptive learning systems	Depends on intended meaning
Adaptive learning system	Algorithms or models that learn from a continual influx of data	System adapts to data
Adaptive performance	How well a human changes their behavior in response to new challenges	Human adapts to task
Adaptive training	Computer systems that can adapt training content and other variables to assist a human with learning a topic	System adapts to human

in addition to which quiz answers were correct and incorrect, an AT system could use eye tracking to determine that the student was not attending to certain portions of the instructional material describing the difference between natural and rational numbers. Then, combining the eye tracking data with the quiz data (observation), the assessment algorithm determines that the student had a misconception with number order of fractions such that they believed a higher denominator meant a bigger number (e.g., 1/10 is bigger than 1/2 because the natural number of 10 is higher on the number line than 2). Finally, the instructional response algorithm determines that the student should review content they skipped over during the lesson but also generates new problems for the student to solve, increasing their complexity as the student moves toward mastery of the topic.

As these examples highlight, both systems are adaptive, but the "intelligence" within the systems are vastly different. When designing AT systems and making decisions for the level of intelligence needed for each AT component, practitioners must consider the nature of the task to be trained and the nature of the data that the system can collect. Knowing these variables provides designers and researchers the resources they need to generate effective adaptive assessment algorithms and consequently how sophisticated those algorithms will need to be to affect the learning experience. Campbell (2014) argued that the level of complexity could vary along all dimensions of the OAR model (see Figures 6.1 and 6.2 for a conceptualization of this idea). All AT systems will most likely have a variety of simple and complex elements, but a system does not need to be complex along every OAR dimension to be effective. Ultimately, AT system designers must use their own judgment along

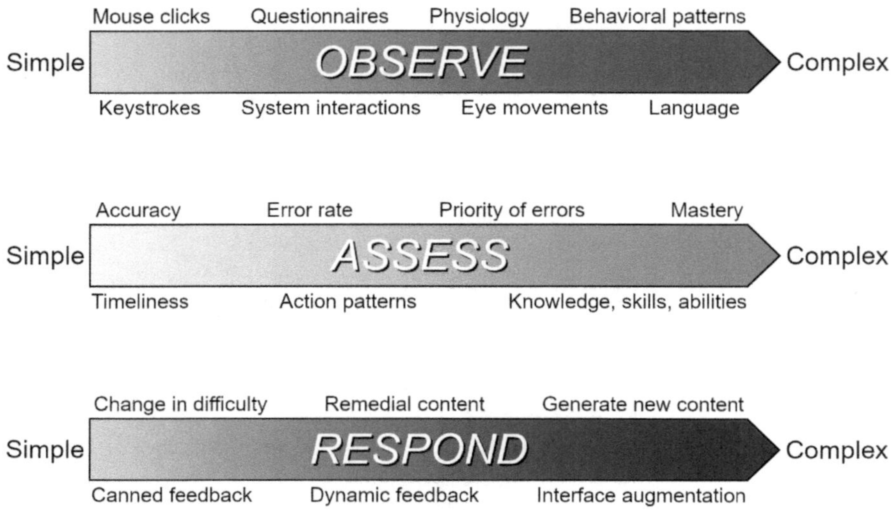

FIGURE 6.1 Elements of the Observe, Assess, Respond (OAR) model. Notional examples are provided along the continuum of simple to complex for each element (adapted from Campbell, 2014).

with evidence from the research literature to implement this model in a valid way. For example, keystrokes may be sufficient for assessing mastery of spelling lessons to provide remedial content, such as phonics lessons. However, in a more complex domain like writing, keystrokes may be too simple to make complex assessments and respond appropriately. An effective training system for writing would need to observe more behavior from the learner, such as elements of language like grammar or semantics.

To be effective, training systems designed under the OAR model should be capable of properly diagnosing the learner's capabilities. This will depend on what data are collected under the "observe" process, which informs how well those data can be assessed to respond in a way that is conducive to learning. Since we are describing adaptive systems, it is understood that learners' capabilities will change over time, and the adaptive system must adjust while the learner continues to gain more knowledge. In the following sections, we offer the concept of mental models as a way of conceptualizing the learning process and continue with theoretical perspectives for assessing the "sweet spot" for learning.

MENTAL MODEL FORMATION DURING THE LEARNING PROCESS

The ultimate goal of training is proper mental model formation. As a concept, mental models emerged out of Craik's (1943) seminal explanation of the human mind and nervous system. He argued that humans were capable of modeling cause-and-effect and broadly that humans could mentally simulate events of the real world. These ideas have been expanded to suggest that mental model construction is an automatic

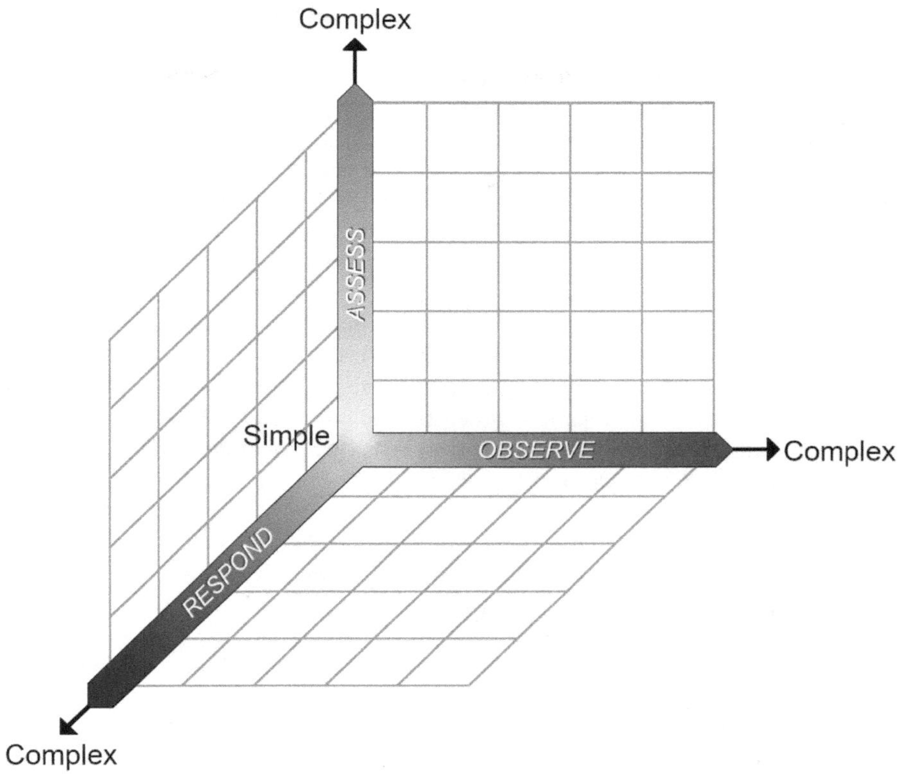

FIGURE 6.2 Three dimensions of OAR. Adaptive training systems can exist anywhere in this three-dimensional space depending on the simplicity or complexity of the observations, assessments, and responses that they can make (adapted from Campbell, 2014).

process that humans engage in to understand the way things work. Johnson-Laird and Byrne (1991) theorized that the construction of mental models was an iterative process. Individuals construct mental models based on their own understanding of how something works, but that model may be revised if alternative information challenges the existing mental model. Importantly, Johnson-Laird and Byrne (1991) contended that humans form mental models semantically – that is, by reasoning with language.

As an example of mental model construction and this iterative process, consider that many people learn about mixing paint colors at a young age. Upon learning this, they will naturally construct a general mental model for mixing colors. Having witnessed that mixing blue paint and yellow paint yields green paint, one could surmise that mixing paint colors together will typically yield a darker color. Continued paint mixing would yield continuously darker colors, until one is left with dark brown or black paint. While observing the causes and effects of mixing colors, the individual forms a mental model to comprehend their observations. This mental model could be described in terms of cause-and-effect: *"the more paint colors that I mix, the darker my resulting paint color will be."* Thus, the learner has constructed a mental model for mixing colors.

However, when mixing colors of light, this previously constructed mental model is incompatible. Combining red light and green light yields yellow light, and continued light mixing would yield continuously lighter colors, until one is left with white light. This may come as a surprise to a young learner who has relied on a mental model for mixing colors when learning with paint. The learner then needs to revise their over-generalized mental model for color mixing to include conditions for what medium is being mixed.

Mental models are not limited to understanding the world through observation, as described above. Mental models can be more sophisticated, allowing the mental simulation of events, motion, and changes over time (Collins & Gentner, 1987). As learners try to understand something new, their mental models undergo initial construction and iterative modification, which can include any of these features (Glaser & Bassok, 1989). Learners begin with loosely structured mental models that must be shaped by undergoing the learning process.

FACILITATING MENTAL MODEL CONSTRUCTION IN ADAPTIVE TRAINING SYSTEMS

In a well-designed AT system, underlying instructional algorithms will change elements of the training to support mental model development based on what the system can observe in the learner's behavior. For example, if an AT system assesses that a learner is struggling with a particular conceptual element of a task, it can respond with tailored feedback to reinforce that under-developed concept.

To illustrate this point, consider a mechanic learning how to replace an engine part. An important concept in engine part replacement is torque. Nuts, bolts, and other fasteners must be torqued with precisely specified amounts of force to ensure engine parts stay attached. If fasteners are under-tightened, a part may eventually become detached due to normal operating vibrations on the engine. If fasteners are over-tightened, they may become damaged such that they no longer fasten the part to the engine effectively. Therefore, meeting proper torque specifications is integral for engine operation. Whether this mechanic is learning from a human instructor or a mixed reality training system, they need a learning intervention on the concept of torque specifications. Where a human instructor might observe the mechanic improperly torquing, an AT system could objectively observe the mechanic's wrench rotations to assess whether they under- or over-torqued. With this objective assessment, the AT system could deliver a remedial lesson, feedback, or other instructional content to ensure the learner understands this concept.

In the example above, the learner may have constructed a mental model that could be described as *"to replace an engine part, I need to attach it to the engine and tighten the bolts that connect it to the engine."* However, without a proper understanding of torque, this mental model is incomplete. After receiving remedial instruction on the concept of torque specifications, this mental model may be reshaped to incorporate mental simulations of the consequences when torque specifications are disregarded.

The AT system detected a deficiency in the mechanic's mental model and *changed the training* to meet their need to refine that mental model. As mentioned, the construction of mental models is an iterative process, but these iterations are well-suited to the iterative learning interventions that are possible with AT.

HOW TO DESIGN ADAPTIVE TRAINING ALGORITHMS TO FIT THE LEARNER'S NEEDS

The goal of an AT system is to adjust any element of the training environment to put the trainee in the "sweet spot" for learning. In essence, the "sweet spot" is a figurative place where a learner faces some challenge but will still have a good likelihood of succeeding. Finding the "sweet spot" ensures learners are not bored with lessons that are too simple and unengaging, but they still must expend effort to proceed through the training. Importantly, the lessons must not be so challenging that they cause the learner to disengage from feeling overwhelmed. This idea may seem simple enough, but complexity grows when considering that the "sweet spot" is going to be different for every learner, and it will change throughout the learning process. A variety of theoretical perspectives suggests the "sweet spot" creates the best opportunity for learners to get the most out of their training. In the following sections, we describe a few perspectives worth considering from the psychological literature, as well as perspectives that have been applied in practical AT research using the OAR paradigm.

ZONE OF PROXIMAL DEVELOPMENT AND THE "SWEET SPOT" PHILOSOPHY

Education-based researchers will be familiar with Vygotsky's concept of the Zone of Proximal Development (ZPD; Vygotsky, 1978). This concept is based on theories of children's learning but posits that the key to learning is understanding what students can accomplish on their own and what students can accomplish with help from another (e.g., skilled peer or teacher). This "zone" between these is considered the ZPD. Vygotsky argued that a learner receiving assistance while in their ZPD would help them consolidate their learning. Essentially, as students accomplish their learning tasks while receiving help, they begin to acquire the skills that enable them to accomplish those same tasks on their own.

Over time, this process "shifts" the ZPD toward more complex learning tasks as the student continues to learn. In AT contexts, this is commonly applied to difficulty adaptation or scaffolding (Goldberg et al., 2015; Sottilare & Brawner, 2021; Van Buskirk et al., 2019), where the difficulty of the training is matched to the learner's capability, or scaffolding is provided when the lesson content changes or adds complexity. With difficulty adaptation as an example (illustration provided in Figure 6.3), training that is too easily passed (determined through observation of performance variables) runs the risk of being too boring for the learner, causing them to disengage with the learning material. This is contrary to training that is too difficult and overwhelms the learner who then gives up. When providing an instructional response such as scaffolding, an AT system could offer support if it assesses that a learner is

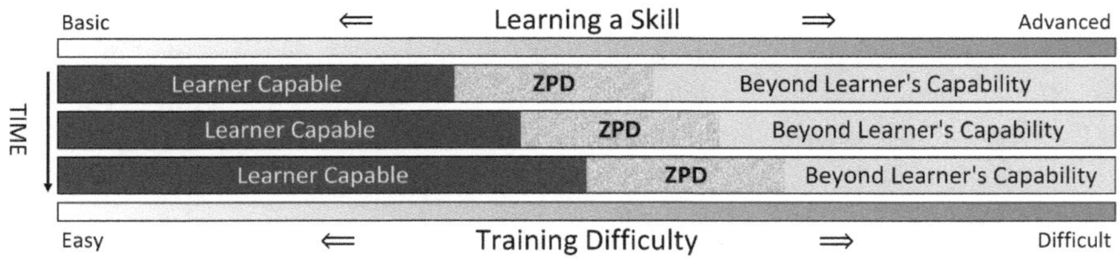

FIGURE 6.3 An illustration of the concept of Zone of Proximal Development (ZPD) as imagined for learning a skill with an adaptive training system, where difficulty adapts over time as the learner's capability increases over time. Providing training to a learner within their ZPD is argued to facilitate more efficient learning than providing training below or above their ZPD.

starting to falter or when the learner proceeds to a more complex lesson. The challenge facing AT systems is to detect when learners have entered these states (observe and assess) and alter the difficulty or the assistance of the training to return to the learner's ZPD (instructional response).

Some researchers view Vygotsky's ZPD as controversial for learning applications (Chaiklin, 2003; Dunn & Lantolf, 1998), since its original focus was on understanding how children develop mentally through different phases of childhood, and they contend that it should not be extended to adult learning. Despite this, the concept of ZPD has played an influential role in science of learning research for decades (Gredler, 2012; Nyikos & Hashimoto, 1997; Puntambekar, 2022; Salomon et al., 1989; Verenikina, 2003). Ultimately, the concept of ZPD as it is used in contemporary research may be too vague to be of use to AT system designers and researchers. For example, an AT system designer can target the learner's ZPD by changing difficulty or modifying the level of scaffolding present in the learning scenario. However, the ZPD perspective does not offer specific guidance where a researcher or designer could predict when a learner is in their ZPD and when they should be pushed for a higher challenge. This is left to the AT system designer to determine with their best judgment, perhaps with the support of prior performance data or subject-matter expertise. Shortcomings aside, ZPD is an easily understood philosophy for many different audiences. However, there are alternative theoretical perspectives that can apply to AT designs that maintain the philosophy of targeting the "sweet spot" for learning.

COGNITIVE LOAD THEORY

Cognitive Load Theory (CLT; Sweller et al., 2011) is another theoretical perspective that has been used in the design and evaluation of AT systems (Marraffino et al., 2021). CLT establishes that learners experience three different types of cognitive load: germane, intrinsic, and extraneous. Germane cognitive load is effortful processing that learners engage in as they are constructing mental models of the learning task. Essentially, learners use their working memory to process the task and commit the information to long-term memory. Intrinsic cognitive load is the inherent tendency of the subject matter to induce load in the learner. This is either a property of the task itself or a function of the learner's knowledge. For example, learning subtraction has a lower intrinsic load than learning division, but the level of intrinsic load induced by division would be greater for an elementary school student than a college mathematics professor. Extraneous cognitive load is any variable not directly relevant to the learning experience that might interfere with cognitive processing (e.g., noise or distractions in the learning environment, poor system usability, and poorly designed learning material). CLT also posits that these different types of load are additive, and each human has a limited capacity for overall cognitive load. If these types of load become too excessive, it can result in cognitive overload, which is undesirable for the learner and learning outcomes (see Figure 6.4 for examples of

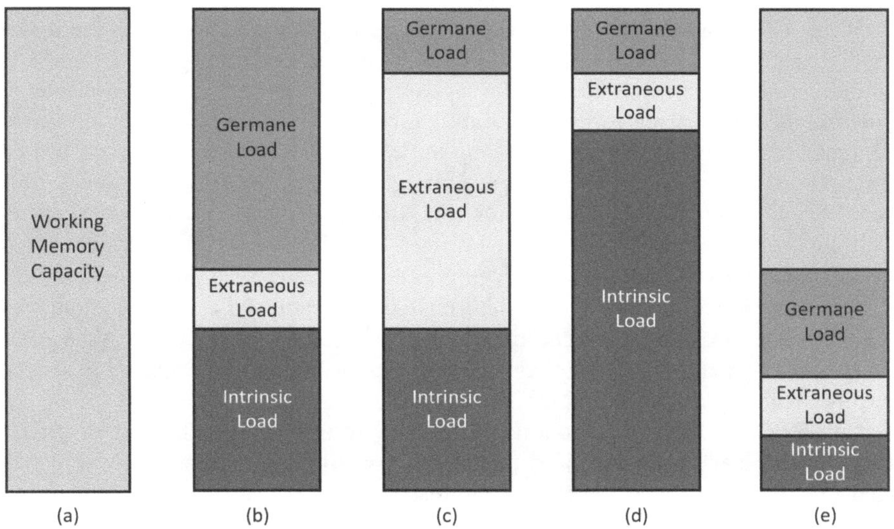

FIGURE 6.4 An example of different learning circumstances as understood through Cognitive Load Theory (CLT). (a) CLT assumes that human working memory capacity is limited. (b) An example of an ideal learning circumstance, where intrinsic load is manageable and extraneous load is minimal, so the learner can dedicate the rest of their attentional capacity to germane load. (c) An example of a poorly designed learning circumstance, where intrinsic load is manageable, but extraneous load is excessive and limits the learner's capacity for germane load. This scenario would likely lead to overload for the learner. (d) An example of a learning circumstance that is beyond the learner's current capability. The intrinsic load is too great for the learner, where even minimal levels of extraneous load will leave the learner with little capacity for germane load. An adaptive intervention would be necessary to reduce the intrinsic load on the learner to avoid overload. (e) An example of a learning circumstance that is beneath the learner's current capability. The intrinsic load is too low for the learner and therefore does not require a substantial capacity for germane load. This task is likely too easy for the learner and may represent a learning inefficiency. An adaptive intervention would be necessary to increase intrinsic load on the learner so that they undergo more germane load to process the learning material.

how to understand different learning circumstances with the assumptions of CLT). However, as learners construct and revise their mental models through germane processing, the intrinsic load they experience will decrease. This facilitates more efficient use of their limited working memory capacity. Ideal learning environments will foster germane load, minimize extraneous load, and manage intrinsic load.

Different types of cognitive load will be highly variable based on the individual. Intrinsic load will impose different demands on individuals based on their own cognitive ability or prior knowledge of the learning content. Extraneous load can vary based on an individual's ability to acclimate to things such as noisy environments or their patience with poor user interface design. These individual differences will mediate the learner's ability to execute germane cognitive processing while learning. Careful identification of the appropriate individual differences (such as prior knowledge or relevant cognitive abilities) for the task would be necessary to support the learner (e.g., the individual differences principle; Mayer, 2009).

When designing an AT system, extraneous load should be reduced to the maximum extent possible to minimize the possibility of overload. Designers are empowered to "design out" extraneous load wherever possible and can manage the level of intrinsic load the learner experiences through performance assessment and training content adaptation. In essence, managing intrinsic load is how an AT system would locate the "sweet spot" for the learner, where germane load can be prioritized. This could be accomplished with something like difficulty adaptation, as more difficult training scenarios will have a higher intrinsic load than less difficult training scenarios. Alternatively, it could be accomplished by lesson selection, where an AT system diagnoses fundamental errors (through observation and assessment) and provides remedial learning interventions (the instructional response). Just as with ZPD, this could also manifest as adaptively supplying scaffolding to the learner to reduce intrinsic load of the subject matter.

Like ZPD, CLT is not without its critics. One of the greatest challenges of CLT is determining and validating how to observe or measure cognitive load. Subjective measures exist but have limitations (Ayres, 2017). Physiological indices of cognitive load have also been tested (Coyne et al., 2009; Haapalainen et al., 2010; Hughes et al., 2019), but these generally assess "cognitive workload" in sum, not the different types of cognitive load described above. Despite CLT's theoretical limitations, efforts continue to identify appropriate physiological markers (Ayres et al., 2021) and develop questionnaires (Krieglstein et al., 2023) to measure the different types of cognitive load.

For research applications, CLT offers a meaningful theoretical framework to make predictions and observations in AT experiments, but the specificity issues with the load types in CLT precludes rigorous experimentation. This should change as physiological and subjective measurement methodology research continues to evolve. For now, researchers can easily manipulate intrinsic and extraneous load but must assume germane load has occurred with indirect post-task measures of learning or performance. However, the design implications and accessibility of CLT are still useful for practitioners: reduce or eliminate extraneous load and adapt the intrinsic load to maximize the potential for germane cognitive processing during learning.

HUMAN PERFORMANCE UNDER STRESS

Yet another perspective for locating the "sweet spot" can be considered from the domain of stress research with human performance (Hancock & Warm, 1989). Their "Trinity of Stress" is structured in the flow of inputs, adaptations, and outputs. This perspective argues that the task itself is the source of stress on the individual (input), the individual must adapt to this stress to continue performing the task (adaptation), and their performance can change because of this process (output). It is important to note that the individual's adaptations to the stress source can be psychological, physiological, or both, and may be deliberate or involuntary.

Considering stress as a continuum from hypostress to hyperstress, Hancock and Warm posited several "zones" for human performance (see Figure 6.5). At the center of this spectrum is the "normative zone" of the individual, which triggers no adaptation on the part of the individual. Hancock and Warm suggested that this zone was transitory in cases of changing task demands. Outside this is the "comfort zone" of the individual, where the input stressors do cause the individual to adapt to the stress of the environment. Surrounding the comfort zone is a zone of psychological limit, which is further surrounded by a zone of physiological limit. Outside of these are zones of dynamic instability where failure is inevitable. Hancock and Warm (1989) argued that performance should decrease as individuals break out of the comfort zone and into the zones of psychological or physiological limits.

This theoretical perspective contributes an additional element of nuance worth considering for designing AT systems. For the "input" aspect of the model, perception of the stress of the environment or task can vary by individual (Matthews & Campbell, 1998). Following that, the "adaptation" aspect accounts for individual differences in how learners cope with the stress of the task. This will vary with individuals from trait and state levels (Schroeder et al., 2019). Both perceptions of the stress and consequent coping can potentially moderate and mediate the "output" (e.g., learning outcomes or performance measures). With this in mind, human performance researchers should account for individual differences in task appraisal and adaptive processes (Matthews, 2016). This is different from individual differences

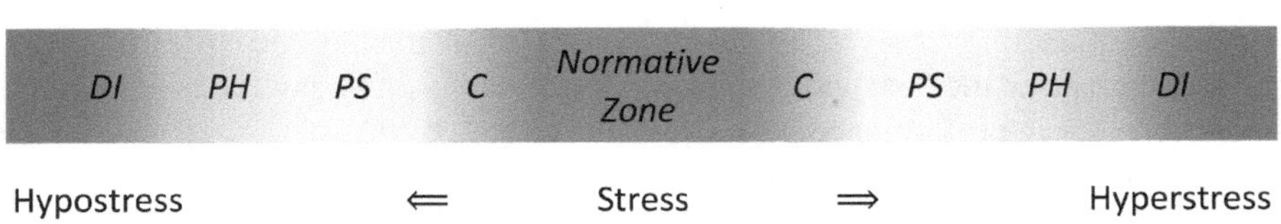

FIGURE 6.5 A simplified adaptation of the Hancock and Warm's (1989) model of performance under stress. As input stress (e.g., task demands) changes, performers will adapt to these changes and be pushed further along this continuum. Each zone represents a greater burden of adaptivity required of the individual, to the point that excessive or insufficient input stress will result in failure (upon reaching the zone of dynamic instability). Performance is predicted to falter the further an individual is away from their normative zone; however, the normative zone can change over time due to training or increasing familiarity with the task.

Note: C = Comfort Zone, PS = Psychological Zone of Maximal Adaptability, PH = Physiological Zone of Maximal Adaptability, DI = Dynamic Instability.

mentioned in previous sections such as prior knowledge and cognitive ability, as the relationship between task-related individual difference measures and performance tends to change as one learns the task to be performed (Schroeder et al., 2019). Hancock and Warm's (1989) model acknowledges how input stress affects the adaptability of individuals but does not offer predictions regarding these relationships. Although this adds complexity to using this model for research purposes, complementary theories of stress and coping with task stress (Lazarus & Folkman, 1984; Matthews & Campbell, 1998) can address this need.

Consistent with the theme of finding the "sweet spot" for learning, AT systems should avoid over-stressing learners such that their level of overwhelm leads to failure (like subjecting a novice to a realistic air traffic control simulation) or under-stressing learners such that their level of boredom leads to failure (such as a slow, boring, extremely low event-rate signal detection task). However, this model expands further to delineate the zones of psychological processing and physiological capability, acknowledging that stressors can come in the form of physical activity in addition to cognitive processing. For complex tasks with physical and mental components, it is important to consider how both independently contribute to the experience of stress in the learner and therefore to their performance (such as the physical demands of learning with gestures in virtual reality, Johnson et al., 2023).

For AT purposes, such a perspective may be particularly useful for dynamic tasks where high performance is required under pressure, such as many military tasks or real-world job tasks. Specifically, Hancock and Warm (1989) argued their model applies to attention-demanding human performance tasks. For the AT field, observation and assessment algorithms may determine a learner's performance is high, which might trigger the instructional response algorithms to increase the difficulty of the next lesson or scenario. However, if the learner is achieving this level of performance under high levels of stress, increasing their difficulty may push them toward a zone of dynamic instability through burnout. A more appropriate adaptation algorithm could be to account for both performance and stress and to hold off on an instructional response and wait until a learner's stress level decreases while performance is maintained. Previous research suggests that decreases in stress naturally occur with additional training or practice (Driskell et al., 2008; Mackworth, 1946). However, this example only accounts for the "input" element of the trinity of stress.

Accounting for the "adaptation" aspect of the trinity of stress can explain additional variance in performance. Variables such as how the learner is coping with the stressful demands of the task can mediate performance outcomes if the learner is coping maladaptively (Matthews, 2002; Matthews & Campbell, 1998; Van Buskirk et al., 2023). Such a variable can be particularly informative for adaptive instructional response algorithms, as previous research suggests that changes in difficulty can induce stress in learners (Cox-Fuenzalida, 2007). Therefore, perhaps another instructional strategy or response should be delivered. Research using this perspective with AT is relatively new compared to ZPD and CLT (Hancock et al., 2024; Schroeder et al., 2019, 2024), but further research is needed to understand how training adaptations (such as dynamic changes in difficulty or adaptive scaffolding) influence the compensatory processes of the learner (such as coping), which further impact performance outcomes.

Unlike the other theoretical models, baseline and post-task measures of stress (either physiological or psychometric) can help determine to what extent a task increases or decreases a learner's perceived stress. However, researchers and practitioners should know that the bounds between the normative zone and the psychological zone of maximal adaptability will vary with different tasks and with individual differences. Hancock and Warm (1989) also suggested that these zones may not necessarily follow a linear continuum but may have discrete thresholds where performance or the learner's adaptability drastically changes.

SUMMARY OF THEORETICAL PERSPECTIVES AND THE IMPORTANCE OF INDIVIDUAL DIFFERENCES

We have presented three different theoretical perspectives for finding the "sweet spot" in AT and provided brief examples of when algorithms can act in accordance with those perspectives. ZPD, CLT, and the stress perspectives offer increasing levels of complexity. Researchers may prefer one model over the other depending on the level of complexity of their hypotheses, the type of task to be trained, or the nature of predictions to be made (i.e., relating to the individual, the task, or both). Researchers should also consider which relevant variables they need to collect prior to selection of a theoretical perspective. This will determine how an AT system can observe learners' behavior, assess that behavior, and respond with an adaptive intervention.

Similarly, designers may find one perspective to be more appropriate than another for creating adaptive instructional content. ZPD and CLT may be more relevant for tasks such as learning mathematical operations, where mastery of concepts or intrinsic load can be operationally well defined. The stress perspective may be more beneficial for training air traffic control tasks that are more complicated and dynamic. There are also other use cases of AT where learning is not the main goal, such as fitness training, where the stress perspective is more appropriate. Ultimately, there is no one-size-fits-all approach to finding a learner's "sweet spot" – this will vary depending on their perceptions of the task and task-relevant individual difference attributes. Selecting a theoretical perspective may also influence which types of responses an adaptive system will execute. ZPD is well-suited to responses such as changes in difficulty, CLT may be better suited to dynamic feedback, and the stress perspective may be better suited to stress regulation interventions.

In each section, we suggested a variety of individual difference variables and how they might influence performance from the lens of each theoretical perspective. Researchers must carefully balance the need for measuring individual differences against overloading their participants and inducing measurement fatigue. There is practically no end to the number of psychometric variables that could be selected, so researchers must carefully identify which ones they expect to be most relevant to the task to be learned. For example, previous research indicates that cognitive abilities such as spatial ability can moderate the effect of learning a spatial task in virtual reality (Johnson et al., 2022). Other research has found that, for learners who tend to use maladaptive stress coping, adapting training in real time can exacerbate distress and workload, which impairs performance (Schroeder et al., 2019).

However, individual differences do not always need to be measured with a questionnaire. Some individual differences can be inferred with behavioral data that an AT system will naturally collect as the learner is using the system. For example, cognitive ability, such as processing speed, could be inferred from data such as how quickly a learner reads a vignette or responds to stimuli. Other research suggests that behavioral data such as redundant actions (Schroeder et al., 2016) and response latency (Schroeder et al., 2024) are correlated with individual differences in neuroticism and maladaptive coping, respectively. If a system is sufficiently sophisticated, AI could be used to identify which of these behavioral variables are relevant for task performance.

However, assessing behavioral data in this way is not without limitation. The effective use of behavioral data as a proxy for individual difference measures assumes the learner is paying attention to the learning content and is not distracted by extraneous stimuli or their own thoughts. Behavioral data are also likely to be confounded with other individual differences, so adaptive algorithms using them may want to weight behavioral data judiciously if used to infer an individual difference attribute in a learner. With that said, questionnaires face similar limitations. The data collected are only as good as the learner's level of earnest response. Ideally, researchers or designers should have a strong, evidence-based justification for selecting behavioral variables or psychometric measures to incorporate into adaptive algorithms.

THEORY AND PRACTICE COME TOGETHER

No matter which theoretical perspective a researcher or designer uses to investigate AT, all AT systems must at minimum be able to observe, assess, and respond. In the previous sections, we have provided many examples of each of these elements, but those creating their first AT system may wonder how to execute these elements in a way that targets the "sweet spot" and maximizes learning gains.

Consider learning an automotive maintenance procedure as an example. If someone is being trained how to replace a part, there are a number of steps that must be followed in a particular order to complete the procedure successfully. For example, if you are replacing an engine alternator, you cannot remove the alternator as the first step. Typically, you must start with disengaging the electrical system (i.e., by disconnecting the battery or any directly connected electrical cables), then release tension on the alternator belt before removing the alternator. There are many sub-steps associated with these steps (selecting the correct tools, removing the right bolts at the right time, etc.), and replacement generally replicates these same steps in reverse order.

A mechanic training someone how to perform this procedure could show them step-by-step how to complete this procedure, which would be the ideal approach (one-on-one training). An AT system instead might show a step-by-step video of the procedure to the learner and then provide them an interactive simulation to try the procedure on their own. This system can observe many different elements of the learner's behavior in the simulation. Did they select the correct tool? Did they remove the right bolt? Was it removed as the correct step of the procedure? How long did it take them to execute their next step? These observations represent basic facets of performing this procedural task successfully.

However, these observations are meaningless on their own without some kind of assessment. A human instructor might observe their learner picked the incorrect tool or removed the wrong part and conclude that they lack declarative knowledge of the steps of the procedure. If the learner does the wrong step at the wrong time, the human instructor may determine that their pupil knows the proper steps, just not in the correct order. Similarly, if a learner is taking longer than usual to execute the next step, their instructor may assume they are not confident in knowing which step is next. These are all assessments that a human instructor can make about their observations of the learner's actions. They may or may not be correct assessments, but they can be useful for determining what kinds of instructional interventions are appropriate.

Determining what to make of these assessments is where instructors and AT designers make or break the adaptive experience. This is where the instructor or adaptive system responds to a learner's needs. This is where an AT system executes its AI to assist the learner with proper mental model formation. For example, a human instructor may observe their pupil taking much longer than usual to get from one step to the next step in a procedure. If they determine that this means they need help knowing which step is next in the procedure, they might offer verbal cues to remind them what they should be doing. An adaptive system could do this in a much more subtle way, like highlighting the next step that needs to be taken (e.g., highlighting a specific part to be removed). This is important in both cases, as leaving a student alone to practice without any feedback risks them encoding the procedure in the wrong order.

Importantly, responding to learners' needs depends on a multitude of factors, as mentioned repeatedly in previous sections. Nevertheless, with this example, designers can consider the thought process of how observations, assessments, and responses can be determined. Once these factors are determined, designers can begin constructing the algorithms that will fuel their AT system. First, a thorough understanding of the task-to-be-trained is helpful and should be obtained either through documentation, educational material, or through other sources of knowledge, such as subject-matter experts. Second, understanding the training experience (e.g., learning this task for the first time) can provide useful information for which elements of training are going to require more attention. Upon learning the task, training a novice how to do it can be informative for the designer. Last, designers must understand what level of performance is acceptable or demonstrates proficiency. As with understanding the task, documentation or subject-matter experts may be authoritative sources of this information (such as service manual guidelines that suggest it should take 1 hour to replace an alternator).

AT systems designed without the aforementioned requisite information can still be adaptive, but the adaptations may not be meaningful if they are not grounded in a sufficient understanding of the material to be trained, its context, or an understanding of the learner's capabilities and learning needs. For instance, how would AT for an alternator replacement differ from a novice mechanic to a NASCAR mechanic? Novice mechanics may be replacing their own alternator in their spare time on a weekend afternoon, whereas a NASCAR mechanic may need to replace an alternator with haste to get their team's vehicle back in the race. Similarly, novice mechanics may have greater learning needs than a NASCAR mechanic. Adaptations may adjust too far out of the "sweet spot" if they are based on improperly assessed observations or if necessary observations are overlooked. Without sufficient task-specific knowledge, learner assessments, and guidance from learning science theory, designers may unintentionally create algorithms that push learners away from their sweet spot, impose too much intrinsic load at the wrong time, or induce a level of stress incongruent with the target task. For example, a well-designed adaptive algorithm for a novice mechanic may assess that steps in the procedure are being performed out of order and provide feedback on the error describing the negative consequences of performing that step out of order. For the NASCAR mechanic, due to the extreme time pressure involved, real-time, detailed feedback may cause high cognitive load and could cause dynamic instability and lead to failure of getting the car back on the track.

Therefore, having knowledge of the task to be trained allows designers to create AT algorithms that are well informed and should approximate toward the intelligence of a human instructor. However, these have been simple examples of algorithmic interventions for adaptively training a straightforward procedure. This represents a basic form of AI, being that it is modeled after how an expert performs the task (an expert model) and adjusts the training through algorithmic "rules" until the learner is performing in accordance with an expert. In the following section, we speculate on some potential uses with more sophisticated applications of AI in AT systems.

GOING BEYOND RULE-BASED ARTIFICIAL INTELLIGENCE FOR ADAPTIVE TRAINING

AI seeks to develop a computer system that behaves like "an intelligent organism," such as an actual human (Raynor, 1999). To this end, AI systems encompass a spectrum of complexity driving the desired intelligent behavior, ranging from simple conditional logic to more advanced models built from observed data. In the context of AT, systems designed to represent an instructor or expert user's mental model of some task and guide trainees throughout their own mental model construction generally serve to replicate the "gold standard" of a one-on-one human instructor (Durlach, 2012), so they can broadly be considered an application of AI. Indeed, the manner in which such systems adapt difficulty to a given learner's performance, individual differences, or other aspects is typically informed by the body of knowledge of subject-matter experts, particularly in the context of intelligent tutoring systems that seek to "embed" domain knowledge into computer-based training solutions (Burns & Capps, 2013). Along with the encoding of mental model construction, the presentation of the training itself might attempt to replicate or mimic the social qualities of a human instructor, such as through a virtual non-player character (Moreno et al., 2001; Schroeder et al., 2020) that can verbally speak or show emotions in a human-like fashion. Such approaches still fall under the umbrella of AI even when they do not specifically emulate human instructional capabilities, as they are designed to replicate human behaviors.

As mentioned previously, the "intelligence" within the OAR paradigm can vary in complexity for AT systems. Up to this point, we have provided examples of rule-based AT algorithms (e.g., if learner commits an error on part A, then provide feedback message for part A). In general, we refer to AT systems built upon hardcoded rules or heuristics as "rule-based systems" (Hayes-Roth, 1985). More precisely, a rule is some exact, logical encoding of a specific set of circumstances, whereas a heuristic is a less well-defined encoding that serves as an estimation or approximation (Raynor, 1999) of a set of circumstances – a tool that is "useful but need not guarantee success" (Romanycia & Pelletier, 1985). Typically, these rules and heuristics are designed to map observed learner input to performance assessments and potentially to training adaptation responses. For example, in a hypothetical AT system dealing with paint mixing, one such rule might be that "mixing blue paint and yellow paint yields green paint," and a learner who violates this rule through their performance might receive some specific training intervention to improve their mental model formation based on this incorrect action. Rules might also consider the number of mistakes made and how this should relate to adaptivity, such as "answering more than 80% of paint mixing questions correctly should prompt a difficulty increase." In contrast, a heuristic would represent such circumstances in a more general sense, such as "mixing two different paints yields a color different than the input colors." Heuristics are often more appropriate sources of adaptivity for complicated tasks, where it can be challenging to distill correct or incorrect actions into precisely defined conditions (Raynor, 1999). AT systems that incorporate such rules or heuristics seek to "[codify] the problem-solving know-how of human experts" (Hayes-Roth, 1985) and are therefore valuable tools for guiding learners through mental model formation.

However, the spectrum of AI approaches extends beyond formulations based on rules and heuristics; for example, more sophisticated AT solutions may incorporate machine learning (ML) approaches that create models implicitly (Mitchell, 1997) through automated pattern analysis of learner performance data without requiring manual task- and/or individual-specific development, which may provide more personalized and/or effective training. Because of task dependency and relevant individual differences, developing an effective AT system that represents and informs the learner's creation of a particular mental model could become costly (Shute & Zapata-Rivera, 2012), so the ability to build such systems without explicit human intervention is a substantial benefit. Training for some tasks is readily representable as a set of steps, rules, or guidelines that a learner should follow, which are often amenable to direct implementations in computer-based AT systems. However, other tasks might lack an explicit relationship between learner input and desired system output, which might better be handled by building models based on sets of examples (Ayodele, 2010). Other tasks might require creativity or problem solving that is otherwise hard for humans (both subject-matter experts and software developers) to explicitly formulate (Colin et al., 2016); such tasks might benefit from more advanced ML models that instead adapt based on statistical inferences on learner outcomes, where an AT system could choose to provide training content to a given learner that has previously been observed to promote increased performance in similar learners. Additionally, ML approaches are capable of modeling aspects of a task beyond what an expert user or instructor might reasonably be able to detect or formulate, such as identifying subtle patterns across very large sets of learner data or encoding a significantly large body of knowledge (Ayodele, 2010).

Just as an AT system seeks to adapt to a particular individual, so too must the design of the system itself adapt to the task being trained. The application of AI or ML requires some analysis of the specific task, learners, and model under consideration. Not all learning tasks require or benefit from sophisticated adaptation schemes beyond simple rules or heuristics; system designers must determine whether the development costs of these models outweigh the benefits afforded. Moreover, it is important to note that these techniques often perform best as supplemental tools in AT development. As such, rather than treating them as replacements for an instructor or expert user, one must examine when and how they can be used effectively – and when their use is not appropriate. Knowledge from one domain may not transfer to another, and AT systems generally incorporate "narrow AI" that focuses explicitly on the specific task being trained (Adams et al., 2012). In other words, AI and ML are not necessarily general-purpose "black boxes" that can be automatically incorporated into an AT system and lead to benefits in training, and overreliance on the use of such "black boxes" may prevent explainability of results or even lead to less effective outcomes than a more thoughtfully developed solution (Rocha et al., 2012).

First, we cover common use cases of AI and ML in AT system development, highlighting developmental considerations for which these techniques can be used as effective supplements in trainers. In keeping with the spirit of AI, we focus on emulating the capabilities of a knowledgeable human instructor and on specific ways to enhance them. In many cases, this still requires subject-matter expertise and/or domain-specific knowledge. Next, we consider limitations of AI and ML approaches and describe where their use in AT systems may not be as beneficial.

When Is AI/ML Useful?

Compared to simpler rule-based approaches, AT systems backed by more advanced AI and ML techniques offer many capabilities that can potentially improve the observe, assess, and response loop. Broadly, we group these advancements into three major categories:

1. The ability to analyze, model, and predict learner data
2. The ability to present real-time instructional content
3. The ability to provide tailored language to and interpret natural language from learners

Data Analysis, Modeling, and Prediction

A key capability of ML is the analysis of patterns in data to build models describing relationships among data and make predictions about novel input without requiring explicit programming (Sugiyama, 2015). For AT systems, relevant data includes learner performance, preferences, individual differences, and physiological measurements (e.g., stress; Finseth et al., 2021), which could all potentially be considered when tailoring content to a particular learner. As an example, human instructors might note that trainees commonly make a specific mistake when learning a task, so they adjust their training by preparing additional resources related to this mistake or even provide these resources ahead of time to prevent the mistake entirely. ML approaches may be able to perform this step automatically and in a general way across many error sources, without requiring manual analysis. In other words, instead of relying on instructor insight, an automated analysis technique may discover that learner mistakes on one specific aspect of a task strongly correlate with mistakes on another aspect, so the AT system can pre-emptively provide additional instruction on this latter mistake source prior to the learner actually making related mistakes.

Furthermore, automated approaches can be effective at explicitly determining causality within datasets (Bontempi & Flauder, 2015; Huang et al., 2020). In a training context, these may manifest as second- or higher-order error sources, such as a mistake later in mental model formation that is ultimately due to a misunderstanding earlier in the process. Such causality chains may be difficult for human instructors to identify, even with high levels of expertise. ML-based approaches are often better able to perform this pattern analysis across large sets of data than human instructors, such as over many experimental sessions for one learner or across data from many learners, and it may be easier for these models to be updated over time as task demands and training requirements change than for expert users and developers to continuously update an AT system (Ayodele, 2010). While such approaches can be fully automated, they may still benefit from human knowledge, such as through human-labeled model training sets.

In general, rule-based systems must take a potentially limitless domain of possible learner actions and outcomes and categorize them using a finite set of prescribed rules. While subject-matter experts can inform this domain, ML-based AT systems may be able to expand the number of recognized categories, leading to instructional content or feedback more explicitly tailored to each particular user. For example, such a system might be able to generate novel verbal feedback or attentional guidance based on precise actions the learner performed, which may not have been predictable and therefore may have no corresponding predefined rule or heuristic to identify them. Additionally, rules and heuristics may combine in unintuitive ways that ML algorithms are better suited to recognize and codify given a set of learner data. However, there is a risk that ML algorithms could yield unintuitive or spurious relationships among confounds in learner data that may be confusing to the learner. This risk will be discussed in greater detail in a later section.

Real-Time Content

A human instructor can monitor a learner's actions in real time and provide feedback or adapt difficulty based on these actions or performance metrics. While rule-based AT systems often seek to replicate this ability, they may be limited in such real-time analysis capabilities. For example, many such systems are only able to provide training interventions after the learner completes a specific sequence of actions rather than at the start of this sequence, perhaps because this particular chain of events was not considered when the rules or heuristics were developed or because the system is simply only able to recognize the final action in the sequence. This limitation is often present in training systems that adapt based on infrequent learner reports or actions; however, a human instructor could also consider the individual steps taken to produce those reports or other learning-relevant behaviors (such as body language or non-verbal communication). Likewise, a more sophisticated ML model might be able to identify such sequences automatically and therefore present feedback in a timelier manner, which may prevent the learner from making errors and disrupting their mental model formation. To achieve such real-time assessment, AT systems can incorporate techniques such as computer vision (see Szeliski (2022) for a broad overview) or natural language processing (NLP; see Khurana et al. (2023) for a discussion of the state-of-the-art) to analyze a learner's actions or textual/verbal input in real time, making them available for adaptivity and assessment purposes.

In addition to monitoring a learner's actions, human instructors can also dynamically guide the learner's attention, perhaps by verbally relaying instructions or by pointing to specific components of the interface. Often, the goal of such guidance is to maximize the amount of time a learner spends actively solving a problem (i.e., germane load) while minimizing time wasted (i.e., extraneous load) due to errors or other obstacles (Merrill et al., 1995). The interface itself can provide guidance of this form, such as by highlighting relevant interface elements or even dimming irrelevant parts, mimicking the capabilities of a one-on-one tutor. Such capabilities can be extended by integrating them with AI or ML techniques: rather than guiding the learner through the interface using prescribed attentional callouts, such as a canned tutorial phase, this guidance could instead be generated dynamically through these real-time monitoring techniques and present customized interventions to each individual learner.

Language

Human instructors training learners on a particular task can tweak content dynamically to better tailor this instruction to a given learner. While this can be emulated through techniques like difficulty adaptation, with a computer system queuing up harder or easier training content in response to learner performance, a human instructor is further able to adjust the actual presentation of the content – for example, by choosing specific words or phrases to answer questions or address mistakes made by the learner. Using NLP techniques, AT software systems can both understand and generate natural language in real time (Khurana et al., 2023).

Many AT systems provide canned or adaptive feedback statements in response to specific learner actions or outcomes (Landsberg et al., 2012a; Schroeder et al., 2020); instead, they could incorporate task-specific language models that can be used to generate novel text-based or audio feedback in real time. Compared to hardcoded statements, these dynamic messages are advantageous in that they allow for variability in messages and can be made more explicitly relevant to a particular learner. Generally, the domain of feedback or other dynamic messages in an adaptive system training a given task is restricted compared to an entire language, so these systems may not need to incorporate extensive language models for this purpose.

Furthermore, natural language generation (NLG; see Gatt and Krahmer (2018) for a recent survey) techniques can allow AT systems to more intelligently respond to text, speech, or other language-based input provided by learners, emulating the ability of a human instructor to respond with their own textual or verbal output. This could include answering learner questions, providing training assessment results, presenting dynamic feedback, or verbally guiding the learner's attention throughout the interface. Additionally, NLG can assist AT system designers with other forms of content generation beyond feedback messages and real-time question answering, such as the creation of training scenario content. As with other uses of advanced AI and ML approaches, methods for automatically generating such content still benefit largely from extensive input from subject-matter experts.

Potential Limitations of AI/ML

In spite of the potential benefits of advanced AI or ML algorithms for AT, there are limitations and general factors to consider (see Cubric (2020) for a survey). In general, one should consider whether the costs of developing and maintaining a more advanced solution outweigh any advantages afforded. Additionally, it is important to note that many of the aforementioned benefits typically still require significant human knowledge and experience during development or even human intervention during actual system use.

Accordingly, we group potential limitations of these methods into three major categories:

1. Lack of applicability and generalizability
2. High development or maintenance costs
3. Lack of instructor/learner understanding or trust

Applicability and Generalizability

Some learning tasks are simple enough that advanced AI or ML models would confer no appreciable training benefits beyond rules or heuristics. For example, rote memorization tasks can often be trained using flashcards, and even simple difficulty adaptivity that schedules cards based on success rates may be sufficient to train a learner successfully (Whitmer et al., 2021). Likewise, even with more complicated tasks, difficulty adaptivity may prove effective even without real-time capabilities or more intelligent models, such as scenario-based training that selects subsequent training content based solely on an overall performance metric for a given scenario (Landsberg et al., 2012b).

ML models may be limited in their ability to generalize across problems (Adams et al., 2012) or populations. For example, speech recognition training data that predominantly features input from a particular race, gender, or age group may be less accurate when analyzing novel input from other groups (Chakraborty et al., 2021; Howard et al., 2017). While such models can be effective in improving training outcomes, those that function as "black boxes" may provide no explicit insights that can lead to benefits on other problems. Though this is often true even for simpler task-specific AT systems as well, the rules or heuristics used in effective training systems may inspire similar rules for other systems more readily than ML models, such as through the demonstrated effectiveness of adapting scenario difficulty (Landsberg et al., 2012b).

Ill-structured problems – those that lack numerical or realistic algorithmic means of verifying a potential solution (Simon, 1973) – were once considered the "exclusive preserve of human problem solvers" (Newell, 1993). Though advances in AI and ML have improved the ability of computers to consider such problems (Colin et al., 2016), certain problems might be less amenable to automated training approaches, such as those that require creativity or insight to solve, feature large input domains, or ultimately rely on inexact judgments made by human instructors.

In some circumstances, developing ML models for use in AT systems could even result in less effective training than simpler approaches. This is particularly common in cases for which it is difficult to obtain and label high-quality data for use in modeling, such as when only a small population of learners is available. Blindly applying ML approaches to a dataset may lead to issues with overfitting, leading to the identification of patterns that truly do not exist (Dietterich, 1995; Schaffer, 1993). In life-critical systems or domains where human life is otherwise at risk, overreliance on ML can have dangerous consequences (Rudin, 2019). For example, studies involving conversational agents with natural language capabilities in healthcare often do not discuss patient safety outcomes (Laranjo et al., 2018). ML models may make inaccurate predictions, and such errors can reinforce maladaptive behaviors in trainees, leading to poor training at best and injury or death at worst.

Costs

Task-specific ML models may be costly to develop, so AT system designers must carefully consider whether they have sufficient resources to build them. Simpler rule-based AT systems, however, might be comparatively easier and cheaper to design, implement, deploy, and maintain. Additionally, ML models designed to enhance training in one specific task may be limited in their ability to generalize to other tasks or populations (Chakraborty et al., 2021; Howard et al., 2017), potentially leading to increased costs when trying to expand the scope of an AT solution. Depending on the specifics of training a model, such as the difficulties of obtaining and labeling data and the availability of ML practitioners, retraining the model as needs change can also be expensive in terms of time and computational resources, making maintenance challenging. As always, these considerations vary based on the particular requirements of a potential training system, so system designers must carefully evaluate their options.

Obtaining and accurately labeling sufficiently large datasets can be cost prohibitive. The benefits of machine-learning-based AT systems may be negated by the cost of actually acquiring relevant data and training models. When designing an AT system for new learning tasks, such data may not exist. This can be a significant challenge in domains where the population of learners is limited or where data collection is otherwise expensive. Subject-matter experts may be better equipped to design AT systems that reflect typical learner populations in the absence of readily available historical data. Furthermore, the quality of this collected data can have large impacts on the effectiveness of trained ML models, limiting both the effectiveness of such systems and trust in their use. Issues such as model overfitting (Dietterich, 1995; Schaffer, 1993) and the bias-variance tradeoff ("the price to pay for achieving low bias is high variance"; Geman et al., 1992) must be considered carefully both during data collection and model training.

With the advent of large language models (LLMs; Kasneci et al., 2023; see Chang et al. (2024) for a recent survey), it may be appealing to incorporate more human-like speech input or output capabilities in AT systems. However, these techniques can require significant development time and cost. While the domain of inputs such systems must recognize and outputs it must produce may be limited, system designers must still construct high-quality text and audio corpora, train models, and evaluate them for both accuracy and speed. In many cases, having a human-in-the-loop to respond to learner input or provide verbal feedback is sufficient or even preferable (Dahlbäck et al., 1993), especially when natural human dialogue is fundamental to the task being trained. For example, LLMs might struggle with realistically handling disagreements and analyzing visual data, and they are often susceptible to adversarial prompts (Chang et al., 2024).

Understanding and Trust

Instructors themselves may misunderstand the capabilities of AI and ML systems, preventing their effective use. For example, the addition of NLG techniques to an AT system may lead instructors to falsely conclude that these systems are true one-to-one replacements for human instructors and are able to perfectly respond to any verbal or textual learner input. Instructors might also not understand that advanced ML techniques cannot simply be incorporated automatically into training systems – instead, they often require large amounts of actual learner performance data to build models and careful consideration regarding the scope and implementation of desired capabilities (Rocha et al., 2012). Furthermore, the ability of these techniques to analyze learner data and make future predictions may mislead instructors to assume that they are able to intuit more information about a learner's mental state than is actually possible through ML models; for instance, tasks requiring the submission of a series of discrete reports may not expose sufficient information about the learner's thought process to extrapolate beyond what a human instructor could.

While rule-based AT systems may be limited in capability compared to more advanced ML-based approaches, they are generally easier for instructors to understand and use effectively. Though systems that instead adapt based on ML models still target mental model formation in learners, the manner in which they achieve this might now be abstracted beyond what explicit, predefined rules and heuristics more clearly convey. However, models that instead promote transparency, interpretability, and/or explainability of ML predictions or decisions, sometimes referred to as white-box models in contrast to the unknowable inner workings of black-box models, may help to bridge this gap (Gunning & Aha, 2019; Rudin & Radin, 2019). However, this may come at the cost of limiting the effectiveness of the models (though this has been argued to be a myth; see Rudin and Radin, 2019 and Rudin, 2019), and the target audience for understanding these models might be those developing them and not necessarily those using them. For example, explanations in an image-labeling task might indicate which portions of an image were considered relevant for its classification but provide no insight as to why these specific portions actually led to the classification (Rudin, 2019), which may provide understanding to computer vision researchers but not to end users relying on the classifier.

Instructors may also lack trust in advanced AI or ML models (Cubric, 2020). Subject-matter experts understand the intricacies of training and performing a given task and may feel that automated systems are not truly capable of providing relevant instruction, assessment, or feedback to learners. Conversely, there is growing concern among the public that advanced computational agents may automate tasks so well that human jobs are threatened (Huang & Rust, 2018), leading to ethical considerations regarding the development and use of such agents. In either case, it may prove difficult to convey the abilities and advantages of these advanced techniques to instructors with respect to AT systems, especially in domains where effective training solutions are already available.

Additionally, advanced AI or ML approaches may be unintentionally perceived more negatively than a simpler system due to a lack of understanding of capabilities or other expectation mismatches on the part of the learner. The push toward more human-like training systems may inadvertently trigger "uncanny valley" effects (Mori et al., 2012) where the combination of human-like and nonhuman-like characteristics prompts a negative reception from learners. For example, replacing textual feedback interventions with a voice-acted non-player character in an AT system may lead to increased learner frustration if that virtual character provides the same feedback word-for-word multiple times during a scenario, as an actual human instructor would not be expected to do so, but these expectations are likely not present for simpler text-based feedback. A human instructor observing the training environment might be better able to appropriately respond to a learner without interrupting them than an algorithm that prompts a particular piece of feedback at the moment some conditions are met, potentially disrupting the learning process. Additionally, perceived mismatches between a virtual character's appearance and voice might prompt "unease" (Meah & Moore, 2014), pointing to the importance of aligning multimodal cues with user expectations. Learners may also experience increased stress if a system that appears able to respond intelligently to verbal questions is unable to accurately do so on occasion, while systems that have no speech recognition capabilities might not elicit such stress as such intelligence was never assumed and therefore no assumption is violated.

SUMMARY

In the present chapter, we have discussed foundations of AT, defining and disambiguating it from oft-confused terms, and the theoretical perspectives that are useful for designers and researchers. Throughout, we have provided examples of AT design approaches, with a focus on designing rule-based algorithms and considerations for individual differences in the learner and their perceptions of the training. Although simpler forms of AI, rule-based AT algorithms can lead to improved learning outcomes as supported by research literature (Johnson et al., 2019; Marraffino et al., 2021; Schroeder et al., 2020). However, AI technology continues to improve and become more accessible, and we speculated on appropriate applications and their risks for AT applications. Ultimately, we contend that a human instructor or expert's involvement in the design of AT algorithms remains necessary, but AT system designers and researchers should consider where and to what extent AI will fit in their system designs.

ACRONYMS AND ABBREVIATIONS

AI	Artificial Intelligence
AT	Adaptive Training
CLT	Cognitive Load Theory
DoD	Department of Defense
LLM	Large Language Model
ML	Machine Learning
NLG	Natural Language Generation
NLP	Natural Language Processing
OAR	Observe, Assess, Respond
ZPD	Zone of Proximal Development

ACKNOWLEDGMENTS

This work was funded by the Office of Naval Research (N0001422WX01634) under Ms. Natalie Steinhauser. Presentation of this material does not constitute or imply its endorsement, recommendation, or favoring by the U.S. Navy or the Department of Defense (DoD). The opinions of the authors expressed herein do not necessarily state or reflect those of the U.S. Navy or DoD.

REFERENCES

Adams, S., Arel, I., Bach, J., Coop, R., Furlan, R., Goertzel, B., …, & Sowa, J. (2012). Mapping the landscape of human-level artificial general intelligence. *AI Magazine*, *33*(1), 25–42. https://doi.org/10.1609/aimag.v33i1.2322

Allworth, E., & Hesketh, B. (1999). Construct-oriented biodata: Capturing change-related and contextually relevant future performance. *International Journal of Selection and Assessment*, *7*(2), 97–111. https://doi.org/10.1111/1468-2389.00110

Ayodele, T. O. (2010). Machine Learning Overview. In Y. Zhang (Ed.), *New Advances in Machine Learning* (pp. 9–18). InTech. https://doi.org/10.5772/9374

Ayres, P. (2017). Subjective Measures of Cognitive Load: What Can They Reliably Measure? In R. Z. Zheng (Ed.), *Cognitive Load Measurement and Application* (pp. 9–28). Routledge. https://doi.org/10.4324/9781315296258-2

Ayres, P., Lee, J. Y., Paas, F., & van Merriënboer, J. J. (2021). The validity of physiological measures to identify differences in intrinsic cognitive load. *Frontiers in Psychology*, *12*, 702538. https://doi.org/10.3389/fpsyg.2021.702538

Billings, D. R. (2012). Efficacy of adaptive feedback strategies in simulation-based training. *Military Psychology*, *24*(2), 114–133. https://doi.org/10.1080/08995605.2012.672905

Bloom, B. S. (1984). The 2 sigma problem: The search for methods of group instruction as effective as one-to-one tutoring. *Educational Researcher*, *13*(6), 4–16. https://doi.org/10.3102/0013189X013006004

Bontempi, G., & Flauder, M. (2015). From dependency to causality: A machine learning approach. *Journal of Machine Learning Research*, *16*(1), 2437–2457. https://doi.org/10.48550/arXiv.1412.6285

Burns, H. L., & Capps, C. G. (2013). Foundations of Intelligent Tutoring Systems: An Introduction. In M. C. Polson, & J. J. Richardson (Eds.), *Foundations of Intelligent Tutoring Systems* (pp. 1–19). New York, NY: Psychology Press.

Campbell, G. E. (2014). *Adaptive, intelligent training systems: Just how "smart" are they?* [Symposium] Adaptive Training Systems Symposium conducted at the Naval Air Systems Command Fellows Lecture Series, Orlando, FL, United States.

Chaiklin, S. (2003). The Zone of Proximal Development in Vygotsky's Analysis of Learning and Instruction. In A. Kozulin, B. Gindis, V. S. Ageyev, & S. M. Miller (Eds.), *Vygotsky's Educational Theory in Cultural Context* (pp. 39–64). Cambridge University Press. https://doi.org/10.1017/CBO9780511840975.004

Chakraborty, J., Majumder, S., & Menzies, T. (2021, August). Bias in machine learning software: Why? how? what to do? In *Proceedings of the 29th ACM Joint Meeting on European Software Engineering Conference and Symposium on the Foundations of Software Engineering* (pp. 429–440). https://doi.org/10.1145/3468264.3468537

Chang, Y., Wang, X., Wang, J., Wu, Y., Yang, L., Zhu, K., …, & Xie, X. (2024). A survey on evaluation of large language models. *ACM Transactions on Intelligent Systems and Technology*, *15*(3), 1–45. https://doi.org/10.1145/3641289

Colin, T. R., Belpaeme, T., Cangelosi, A., & Hemion, N. (2016). Hierarchical reinforcement learning as creative problem solving. *Robotics and Autonomous Systems*, *86*(C), 196–206. https://doi.org/10.1016/j.robot.2016.08.021

Collins, A., & Gentner, D. (1987). How People Construct Mental Models. In D. Holland, & N. Quinn (Eds.), *Cultural Models in Language and Thought* (pp. 243–265). Cambridge University Press. https://doi.org/10.1017/CBO9780511607660.011

Cox-Fuenzalida, L. E. (2007). Effect of workload history on task performance. *Human Factors*, *49*(2), 277–291. https://doi.org/10.1518/001872007X312496

Coyne, J. T., Baldwin, C., Cole, A., Sibley, C., & Roberts, D. M. (2009). Applying real time physiological measures of cognitive load to improve training. In Foundations of Augmented Cognition. Neuroergonomics and Operational Neuroscience: 5th International Conference, FAC 2009 Held as Part of HCI International 2009 San Diego, CA, USA, July 19–24, 2009 Proceedings 5 (pp. 469–478). Berlin Heidelberg: Springer. https://doi.org/10.1007/978-3-642-02812-0_55

Craik, K. J. W. (1943). *The Nature of Explanation*. Cambridge, UK: Cambridge University Press.

Cronbach, L. J. (1957). The two disciplines of scientific psychology. *American Psychologist*, *12*(11), 671–684. https://doi.org/10.1037/h0043943

Cubric, M. (2020). Drivers, barriers and social considerations for AI adoption in business and management: A tertiary study. *Technology in Society*, *62*, 101257. https://doi.org/10.1016/j.techsoc.2020.101257

Dahlbäck, N., Jönsson, A., & Ahrenberg, L. (1993, February). Wizard of Oz studies: why and how. In *Proceedings of the 1st International Conference on Intelligent User Interfaces* (pp. 193–200). https://doi.org/10.1016/0950-7051(93)90017-N

Dietterich, T. (1995). Overfitting and undercomputing in machine learning. *ACM Computing Surveys*, *27*(3), 326–327. https://doi.org/10.1145/212094.212114

Driskell, J. E., Salas, E., Johnston, J. H., & Wollert, T. N. (2008). Stress Exposure Training: An Event-Based Approach. In P. A. Hancock, & J. L. Szalma (Eds.), *Performance under Stress* (pp. 271–286). Farnham, UK: Ashgate Publishing Company.

Dunn, W. E., & Lantolf, J. P. (1998). Vygotsky's zone of proximal development and Krashen's i+ 1: Incommensurable constructs; Incommensurable theories. *Language Learning*, *48*(3), 411–442. https://doi.org/10.1111/0023-8333.00048

Durlach, P. J. (2012). A Road Ahead for Adaptive Training Technology. In P. J. Durlach, & A. M. Lesgold (Ed.), *Adaptive Technologies for Training and Education* (pp. 331–341). Cambridge University Press. https://dl.acm.org/doi/abs/10.5555/2181146

Finseth, T., Dorneich, M. C., Keren, N., Franke, W., Vardeman, S., Segal, J., …, & Thompson, K. (2021, September). The effectiveness of adaptive training for stress inoculation in a simulated astronaut task. In *Proceedings of the Human Factors and Ergonomics Society Annual Meeting* (Vol. 65, No. 1, pp. 1541–1545). Los Angeles, CA: SAGE Publications. https://doi.org/10.1177/1071181321651241

Fraulini, N. W., Marraffino, M. D., Garibaldi, A. E., Johnson, C. I., & Whitmer, D. E. (2024). Adaptive training instructional interventions: A meta-analysis. *Military Psychology.*

Gatt, A., & Krahmer, E. (2018). Survey of the state of the art in natural language generation: Core tasks, applications and evaluation. *Journal of Artificial Intelligence Research, 61*(2018), 65–170. https://doi.org/10.1613/jair.5477

Geman, S., Bienenstock, E., & Doursat, R. (1992). Neural networks and the bias/variance dilemma. *Neural Computation, 4*(1), 1–58. https://doi.org/10.1162/neco.1992.4.1.1

Glaser, R., & Bassok, M. (1989). Learning theory and the study of instruction. *Annual Review of Psychology, 40*(1), 631–666. https://doi.org/10.1146/annurev.ps.40.020189.003215

Goldberg, B., Sinatra, A., Sottilare, R., Moss, J., & Graesser, A. (2015). Instructional Management for Adaptive Training and Education in Support of the US Army Learning Model: Research Outline. US Army Research Laboratory Special Report (ARL-SR-0345).

Gredler, M. E. (2012). Understanding Vygotsky for the classroom: Is it too late? *Educational Psychology Review, 24*, 113–131. https://doi.org/10.1007/s10648-011-9183-6

Gunning, D., & Aha, D. (2019). DARPA's explainable artificial intelligence (XAI) program. *AI Magazine, 40*(2), 44–58. https://doi.org/10.1609/aimag.v40i2.2850

Haapalainen, E., Kim, S., Forlizzi, J. F., & Dey, A. K. (2010, September). Psycho-physiological measures for assessing cognitive load. In *Proceedings of the 12th ACM International Conference on Ubiquitous Computing* (pp. 301–310). https://doi.org/10.1145/1864349.1864395

Hancock, G. M., Schroeder, B. L., Rivera, J. A., Thayer, S. C., Hochreiter, J. E., Diaz, Y. V., Gruber, M. E., & Van Buskirk, W. L. (2024, July 27). *Adaptive versus maladaptive coping: Correlations with performance in adaptive training.* [Conference Session] *15th International Conference on Applied Human Factors and Ergonomics, Nice, France.*

Hancock, P. A., & Warm, J. S. (1989). A dynamic model of stress and sustained attention. *Human Factors, 31*(5), 519–537. https://doi.org/10.1177/001872088903100503

Hayes-Roth, F. (1985). Rule-based systems. *Communications of the ACM, 28*(9), 921–932. https://doi.org/10.1145/4284.4286

Howard, A., Zhang, C., & Horvitz, E. (2017, March). Addressing bias in machine learning algorithms: A pilot study on emotion recognition for intelligent systems. In *2017 IEEE Workshop on Advanced Robotics and its Social Impacts* (pp. 1–7). https://doi.org/10.1109/ARSO.2017.8025197

Huang, J. L., Ryan, A. M., Zabel, K. L., & Palmer, A. (2014). Personality and adaptive performance at work: A meta-analytic investigation. *Journal of Applied Psychology, 99*(1), 162–179. https://doi.org/10.1037/a0034285

Huang, M. H., & Rust, R. T. (2018). Artificial intelligence in service. *Journal of Service Research, 21*(2), 155–172. https://doi.org/10.1177/1094670517752459

Huang, Y., Fu, Z., & Franzke, C. L. (2020). Detecting causality from time series in a machine learning framework. *Chaos: An Interdisciplinary Journal of Nonlinear Science, 30*(6), 063116. https://doi.org/10.1063/5.0007670

Hughes, A. M., Hancock, G. M., Marlow, S. L., Stowers, K., & Salas, E. (2019). Cardiac measures of cognitive workload: A meta-analysis. *Human Factors, 61*(3), 393–414. https://doi.org/10.1177/0018720819830553

Johnson, C. I., Bailey, S. K., Schroeder, B. L., & Marraffino, M. D. (2022). Procedural learning in virtual reality: The role of immersion, interactivity, and spatial ability. *Technology, Mind, and Behavior, 3*(4), 1–15. https://doi.org/10.1037/tmb0000087

Johnson, C. I., Fraulini, N. W., Peterson, E. K., Entinger, J., & Whitmer, D. E. (2023). Exploring Hand Tracking and Controller-Based Interactions in a VR Object Manipulation Task. In *Proceedings of the International Conference on Human-Computer Interaction* (pp. 64–81). Cham: Springer Nature Switzerland. https://doi.org/10.1007/978-3-031-48050-8_5

Johnson, C. I., Marraffino, M. D., Whitmer, D. E., & Bailey, S. K. (2019). Developing an Adaptive Trainer for Joint Terminal Attack Controllers. In *Proceedings of the International Conference on Human-Computer Interaction* (pp. 314–326). Cham: Springer. https://doi.org/10.1007/978-3-030-22341-0_25

Johnson-Laird, P. N., & Byrne, R. M. (1991). *Deduction.* Hillsdale, NJ: Lawrence Erlbaum Associates, Inc.

Kasneci, E., Seßler, K., Küchemann, S., Bannert, M., Dementieva, D., Fischer, F., …, & Kasneci, G. (2023). ChatGPT for good? On opportunities and challenges of large language models for education. *Learning and Individual Differences, 103*, 102274. https://doi.org/10.1016/j.lindif.2023.102274

Kelley, C. R. (1969). What is adaptive training? *Human Factors, 11*(6), 547–556. https://doi.org/10.1177/001872086901100602

Khurana, D., Koli, A., Khatter, K., & Singh, S. (2023). Natural language processing: State of the art, current trends and challenges. *Multimedia Tools and Applications, 82*(3), 3713–3744. https://doi.org/10.1007/s11042-022-13428-4

Krieglstein, F., Beege, M., Rey, G. D., Sanchez-Stockhammer, C., & Schneider, S. (2023). Development and validation of a theory-based questionnaire to measure different types of cognitive load. *Educational Psychology Review, 35*(1), 1–37. https://doi.org/10.1007/s10648-023-09738-0

Landsberg, C. R., Astwood, R. S. Jr, Van Buskirk, W. L., Townsend, L. N., Steinhauser, N. B., & Mercado, A. D. (2012a). Review of adaptive training system techniques. *Military Psychology, 24*(2), 96–113. https://doi.org/10.1080/08995605.2012.672903

Landsberg, C. R., Bailey, S. K. T., Van Buskirk, W. L., Gonzalez-Holland, E., & Johnson, C. I. (2016). Designing effective feedback in adaptive training systems. In *Proceedings of the Interservice/Industry Training, Simulation & Education Conference* (Vol. 312, pp. 1–12).

Landsberg, C. R., Mercado, A. D., Van Buskirk, W. L., Lineberry, M., & Steinhauser, N. (2012b, September). Evaluation of an adaptive training system for submarine periscope operations. In *Proceedings of the Human Factors and Ergonomics Society Annual Meeting* (Vol. 56, No. 1, pp. 2422–2426). Los Angeles, CA: SAGE Publications. https://doi.org/10.1177/1071181312561493

Laranjo, L., Dunn, A. G., Tong, H. L., Kocaballi, A. B., Chen, J., Bashir, R., …, & Coiera, E. (2018). Conversational agents in healthcare: A systematic review. *Journal of the American Medical Informatics Association, 25*(9), 1248–1258. https://doi.org/10.1093/jamia/ocy072

Lazarus, R. S., & Folkman, S. (1984). *Stress, Appraisal, and Coping.* New York: Springer-Verlag.

Mackworth, N. (1946). Effects of heat on wireless operators hearing and recording Morse code messages. *British Journal of Industrial Medicine, 3*, 143–158.

Marraffino, M. D., Schroeder, B. L., Fraulini, N. W., Van Buskirk, W. L., & Johnson, C. I. (2021). Adapting training in real time: An empirical test of adaptive difficulty schedules. *Military Psychology, 33*(3), 136–151. https://doi.org/10.1080/08995605.2021.1897451

Martin, A. J. (2017). Adaptability—What it is and What it is not: Comment on Chandra and Leong (2016). *American Psychologist, 72*(7), 696–698. https://doi.org/10.1037/amp0000163

Matthews, G. (2002). Towards a transactional ergonomics for driver stress and fatigue. *Theoretical Issues in Ergonomics Science*, *3*(2), 195–211. https://doi.org/10.1080/14639220210124120

Matthews, G. (2016). Multidimensional profiling of task stress states for human factors: A brief review. *Human Factors*, *58*(6), 801–813. https://doi.org/10.1177/0018720816653688

Matthews, G., & Campbell, S. E. (1998, October). Task-induced stress and individual differences in coping. In *Proceedings of the Human Factors and Ergonomics Society Annual Meeting* (Vol. 42, No. 11, pp. 821–825). SAGE Publications. https://doi.org/10.1177/154193129804201111

Mayer, R. E. (2009). *Multimedia learning* (2nd ed.). Cambridge, England: Cambridge University Press.

Meah, L. F., & Moore, R. K. (2014). The uncanny valley: A focus on misaligned cues. In Social Robotics: 6th International Conference, ICSR 2014, Sydney, NSW, Australia, October 27–29, 2014. Proceedings 6 (pp. 256–265). Springer International Publishing. https://doi.org/10.1007/978-3-319-11973-1_26

Merrill, D. C., Reiser, B. J., Merrill, S. K., & Landes, S. (1995). Tutoring: Guided learning by doing. *Cognition and Instruction*, *13*(3), 315–372. https://doi.org/10.1207/s1532690xci1303_1

Mitchell, T. (1997). *Machine Learning*. McGraw Hill. https://dl.acm.org/doi/abs/10.5555/541177

Moreno, R., Mayer, R. E., Spires, H. A., & Lester, J. C. (2001). The case for social agency in computer-based teaching: Do students learn more deeply when they interact with animated pedagogical agents? *Cognition and Instruction*, *19*(2), 177–213. https://doi.org/10.1207/S1532690XCI1902_02

Mori, M., MacDorman, K. F., & Kageki, N. (2012). The uncanny valley [from the field]. *IEEE Robotics & Automation Magazine*, *19*(2), 98–100. https://doi.org/10.1109/MRA.2012.2192811

Nancekivell, S. E., Sun, X., Gelman, S. A., & Shah, P. (2021). A slippery myth: How learning style beliefs shape reasoning about multimodal instruction and related scientific evidence. *Cognitive Science*, *45*(10), e13047. https://doi.org/10.1111/cogs.13047

Newell, A. (1993). *Heuristic Programming: Ill-Structured Problems* (pp. 3–54). Cambridge, MA, USA: MIT Press.

Nyikos, M., & Hashimoto, R. (1997). Constructivist theory applied to collaborative learning in teacher education: In search of ZPD. *The Modern Language Journal*, *81*(4), 506–517. https://doi.org/10.2307/328893

Park, O., & Lee, J. (2004). Adaptive Instructional Systems. In D. H. Jonassen (Ed.), *Handbook of Research for Educational Communications and Technology* (pp. 651–685). Mahwah, NJ: Lawrence Erlbaum.

Pashler, H., McDaniel, M., Rohrer, D., & Bjork, R. (2009). Learning styles: Concepts and evidence. *Psychological Science in the Public Interest*, *9*(3), 105–119. https://doi.org/10.1111/j.1539-6053.2009.01038.x

Pulakos, E. D., Arad, S., Donovan, M. A., & Plamondon, K. E. (2000). Adaptability in the workplace: Development of a taxonomy of adaptive performance. *Journal of Applied Psychology*, *85*(4), 612–624. https://doi.org/10.1037/0021-9010.85.4.612

Puntambekar, S. (2022). Distributed scaffolding: Scaffolding students in classroom environments. *Educational Psychology Review*, *34*(1), 451–472. https://doi.org/10.1007/s10648-021-09636-3

Raynor, W. J. (1999). *The International Dictionary of Artificial Intelligence*. Glenlake Publishing Company. https://dl.acm.org/doi/abs/10.5555/1525542

Rocha, A., Papa, J. P., & Meira, L. A. (2012). How far do we get using machine learning black-boxes? *International Journal of Pattern Recognition and Artificial Intelligence*, *26*(2), 1–24. https://doi.org/10.1142/S0218001412610010

Romanycia, M. H., & Pelletier, F. J. (1985). What is a heuristic? *Computational Intelligence*, *1*(1), 47–58. https://doi.org/10.1111/j.1467-8640.1985.tb00058.x

Rudin, C. (2019). Stop explaining black box machine learning models for high stakes decisions and use interpretable models instead. *Nature Machine Intelligence*, *1*(5), 206–215. https://doi.org/10.1038/s42256-019-0048-x

Rudin, C., & Radin, J. (2019). Why are we using black box models in AI when we don't need to? A lesson from an explainable AI competition. *Harvard Data Science Review*, *1*(2), 1–9. https://doi.org/10.1162/99608f92.5a8a3a3d

Salomon, G., Globerson, T., & Guterman, E. (1989). The computer as a zone of proximal development: Internalizing reading-related metacognitions from a reading partner. *Journal of Educational Psychology*, *81*(4), 620–627. https://doi.org/10.1037/0022-0663.81.4.620

Schaffer, C. (1993). Overfitting avoidance as bias. *Machine Learning*, *10*, 153–178. https://doi.org/10.1007/BF00993504

Schroeder, B. L., Fraulini, N. W., Marraffino, M. D., Van Buskirk, W. L., & Johnson, C. I. (2019, November). Individual differences in adaptive training: Distress, workload, and coping with changes in difficulty. In *Proceedings of the Human Factors and Ergonomics Society Annual Meeting* (Vol. 63, No. 1, pp. 2154–2155). Los Angeles, CA: SAGE Publications. https://doi.org/10.1177/1071181319631033

Schroeder, B. L., Fraulini, N. W., Van Buskirk, W. L., & Johnson, C. I. (2020). Using a Non-Player Character to Improve Training Outcomes for Submarine Electronic Warfare Operators. In R. Sottilare, & J. Schwarz (Eds.), *Adaptive Instructional Systems. HCII 2020. Lecture Notes in Computer Science* (Vol. 12214, pp. 531–542). Cham: Springer. https://doi.org/10.1007/978-3-030-50788-6_39

Schroeder, B. L., Leyva, K., Stowers, K., Lewis, J. E., & Sims, V. K. (2016, September). Investigating usability, user preferences, ergonomics, and player performance in StarCraft II. In *Proceedings of the Human Factors and Ergonomics Society Annual Meeting* (Vol. 60, No. 1, pp. 1210–1214). Los Angeles, CA: SAGE Publications. https://doi.org/10.1177/1541931213601283

Schroeder, B. L., Van Buskirk, W. L., Aros, M., Hochreiter, J. E., & Fraulini, N. W. (2023, July). Which is better individualized training for a novel, complex task? Learner control vs. feedback algorithms. In *Proceedings of the International Conference on Human-Computer Interaction* (pp. 236–252). Cham: Springer Nature Switzerland. https://doi.org/10.1007/978-3-031-34735-1_17

Schroeder, B. L., Van Buskirk, W. L., Hochreiter, J. E., & Hancock, G. M. (2024, July). Regulating stress in complex tasks: Human performance implications of adaptive and maladaptive coping strategies. In *Proceedings of the International Conference on Human-Computer Interaction* (pp. 189–203). Cham: Springer Nature Switzerland. https://doi.org/10.1007/978-3-031-60609-0_14

Shute, V. J., & Zapata-Rivera, D. (2012). Adaptive educational systems. *Adaptive Technologies for Training and Education*, *7*(27), 1–35.

Simon, H. A. (1973). The structure of ill structured problems. *Artificial Intelligence*, *4*(3), 181–201. https://doi.org/10.1016/0004-3702(73)90011-8

Sottilare, R. A., & Brawner, K. W. (2021, July). Scaling adaptive instructional system (AIS) architectures in low-adaptive training ecosystems. In *Proceedings of the International Conference on Human-Computer Interaction* (pp. 298–310). Cham: Springer International Publishing. https://doi.org/10.1007/978-3-030-77857-6_20

Sugiyama, M. (2015). *Introduction to Statistical Machine Learning*. Burlington, MA: Morgan Kaufmann.

Sweller, J., Kalyuga, S., & Ayres, P. (2011). In J. M. Spector, & S. P. Lajoie (Eds.), *Cognitive Load Theory*. New York: Springer. https://doi.org/10.1007%2F978-1-4419-8126-4

Szeliski, R. (2022). *Computer Vision: Algorithms and Applications*. Springer Nature. https://doi.org/10.1007%2F978-1-84882-935-0

Van Buskirk, W. L., Fraulini, N. W., Schroeder, B. L., Johnson, C. I., & Marraffino, M. D. (2019). Application of theory to the development of an adaptive training system for a submarine electronic warfare task. In *Proceedings of the International Conference on Human Computer Interaction* (pp. 352–362). Cham, Switzerland: Springer International Publishing. https://doi.org/10.1007/978-3-030-22341-0_28

Van Buskirk, W. L., Schroeder, B. L., Aros, M., & Hochreiter, J. E. (2023, July). Stress and coping with task difficulty: Investigating the utility of a micro-adaptive aptitude treatment interaction approach. In *Proceedings of the International Conference on Human-Computer Interaction* (pp. 253–264). Cham: Springer Nature Switzerland. https://doi.org/10.1007/978-3-031-34735-1_18

van Dam, K., & Meulders, M. (2021). The adaptability scale: Development, internal consistency, and initial validity evidence. *European Journal of Psychological Assessment, 37*(2), 123–134. https://doi.org/10.1027/1015-5759/a000591

VanLehn, K. (2011). The relative effectiveness of human tutoring, intelligent tutoring systems, and other tutoring systems. *Educational Psychologist, 46*(4), 197–221. https://doi.org/10.1080/00461520.2011.611369

Verenikina, I. (2003). *Understanding Scaffolding and the ZPD in Educational Research.* University of Wollongong Australia. https://ro.uow.edu.au/edupapers/381

Vygotsky, L. S. (1978). *Mind in Society: The Development of Higher Psychological Processes.* Cambridge, MA: Harvard University Press.

Whitmer, D. E., Johnson, C. I., Marraffino, M. D., & Hovorka, J. (2021, July). Using adaptive flashcards for automotive maintenance training in the wild. In *Proceedings of the International Conference on Human-Computer Interaction* (pp. 466–480). Cham: Springer International Publishing. https://doi.org/10.1007/978-3-030-77857-6_33

Wickens, C. D., & Hollands, J. (2000). *Engineering Psychology and Human Performance* (3rd ed.). Upper Saddle River, NJ: Prentice Hall.

Zliobaite, I., Bifet, A., Gaber, M., Gabrys, B., Gama, J., Minku, L., & Musial, K. (2012). Next challenges for adaptive learning systems. *Association for Computing Machinery SIGKDD Explorations Newsletter, 14*(1), 48–55. https://doi.org/10.1145/2408736.2408746

7 From Manual to Machine Learning
Reflecting on the Development of an Adaptive Training System for a Military Decision-Making Task

Cheryl I. Johnson, Matthew D. Marraffino, and Jason E. Hochreiter

FROM MANUAL TO MACHINE LEARNING: REFLECTING ON THE DEVELOPMENT OF AN ADAPTIVE TRAINING SYSTEM FOR A MILITARY DECISION-MAKING TASK

The U.S. military is committed to reforming training and education across the services to break away from traditional sage-on-the-stage, one-size-fits-all teaching approaches and provide more modern guide-on-the-side, learner-centered approaches. These modern approaches are designed to take advantage of technology-based training solutions that incorporate individualized lessons, promote critical thinking, and provide the right training at the right time. These reforms are described in recent publications from military leaders, including the Chief of Naval Operations' Navigation Plan (Gilday, 2022), the U.S. Marine Corps (USMC) Commandant's Planning Guidance (Berger, 2019), and the Army's Learning Concept for Training and Education 2020–2040 (Training and Doctrine Command, 2017). One particular technology that can meet the military's training modernization needs is adaptive training.

Adaptive training is a computer-based training solution designed to simulate one-on-one tutoring by adjusting instruction in response to an individual's performance, ability, or some other characteristics (Landsberg, Astwood et al., 2012; Park & Lee, 2004; Shute & Zapata-Rivera, 2012). Adaptive training approaches are attractive to military audiences because they have been shown to increase learning outcomes and optimize limited classroom time with instructors by providing students personalized training that does not necessarily require dedicated class time and resources (Barto et al., 2020; Bond et al., 2019; Landsberg, Mercado et al., 2012; Marraffino et al., 2019; Van Buskirk et al., 2019; Whitmer et al., 2021).

 DOI: 10.1201/9781032701639-7

Despite these clear benefits, deploying adaptive training solutions at scale has proven challenging. For example, adaptive training systems are time-consuming and costly to develop because of the domain expertise required to develop appropriate content, assessments, and tailored instructional interventions in addition to the programming expertise typically needed to build the system. As a result, much of the investment in this space has been funded by Department of Defense research organizations to meet the needs of particular audiences while also serving as use cases to answer important scientific questions, and it is not usually funded directly by the military population who needs the training. However, recent advancements in artificial intelligence (AI) and machine learning (ML) may help alleviate some of the time and cost associated with adaptive training system development and maintenance by assisting in rapid content creation, student assessment, and feedback generation. Perhaps then adaptive training can truly meet its full potential for military training and education.

In this chapter, we describe our approach to developing the Adaptive Trainer for Terminal Attack Controllers (ATTAC), which is an adaptive scenario-based training system for a critical planning phase of close air support (CAS) missions. Specifically, ATTAC adapts scenario difficulty and feedback based on an individual trainee's performance during a training episode to provide Joint Terminal Attack Controllers (JTACs) with reps and sets in realistic, complex decision-making situations. In the following sections, we explain more about the task and how ATTAC works, and we share results from two training effectiveness evaluations of ATTAC conducted with USMC students. Next, we discuss some of the unique challenges we met when developing ATTAC and explore how AI/ML techniques could be applied to speed up the development process, refine the adaptive algorithms as more data are collected from trainees, and help to manage the maintenance of ATTAC to stay current as new tactics, equipment, and doctrine become available. Lastly, we recognize that trust is a critical element in acceptance of AI/ML-assisted technologies, and we offer suggestions that may increase instructors' feelings of trust. Overall, this chapter offers a retrospective on the painstakingly manual process we used when developing ATTAC and how we would approach its development differently now that AI/ML techniques are more readily available and approachable. We hope that our reflections here may be helpful to other researchers and system developers as they build adaptive training solutions in the future.

WHAT IS CLOSE AIR SUPPORT AND THE ROLE OF THE JOINT TERMINAL ATTACK CONTROLLER?

To understand ATTAC's training goals, it is helpful to start with an explanation of the task and the role of the JTAC. CAS missions are ground strikes on hostile targets carried out by aircraft that occur within close proximity of friendly forces. JTACs are certified service members responsible for directing the actions of the attacking aircraft. Their role is critical in coordinating between ground forces and aircraft to ensure the safe and effective delivery of firepower to meet the commander's intent for the mission. The JTAC's role in CAS is dynamic and involves communicating with both ground units and aircraft pilots, planning the attack, and providing detailed target information and clearance for airstrikes.

Executing a CAS mission involves a complex 12-step process. Game plan development is just one of those steps, but it is a critical one that sets the stage for the whole CAS mission. During game plan development, the JTAC coordinates with the attacking aircraft to determine four key elements of the attack: Type, Method, Ordnance, and Interval (TMOI). The Type of attack (i.e., Type 1, 2, or 3) refers to the amount of control the JTAC requires over the attack. The Method of attack (i.e., bomb on coordinate or bomb on target) refers to how the JTAC and aircraft will correlate the target to confirm they are referencing the same target. The Ordnance is the aircraft weapon that will be deployed on the target, and ordnance selection should be based on its ability to be effective and prosecute the target safely given other contextual factors in the mission. Finally, the Interval is the amount of time separation the JTAC requires between subsequent attacks by a section of two aircraft. It is important to note that sometimes the Type, Method, and Ordnance decisions will be the same for both aircraft executing the attack, but with more complicated game plans, these decisions may be different for the two aircraft.

Although the game plan is composed of only four to seven elements (TMO for each aircraft plus Interval), the complex interaction between them creates a complicated decision-making process. In addition to the fact that in some situations there is more than one effective approach, game plan elements cannot be considered in isolation and instead must be treated holistically. This can lead to situations in which changing just one element of an effective game plan turns it to one that is no longer effective. Given the complexities and nuance to this aspect of CAS, JTAC instructors indicated that students tend to struggle while learning this process and would benefit from additional practice.

WHAT IS ATTAC AND HOW DOES IT WORK?

ATTAC is a standalone adaptive scenario-based trainer that provides trainees with reps and sets on CAS game plan development. ATTAC works by presenting scenarios for JTACs to develop an appropriate game plan by selecting the individual game plan elements (i.e., TMOI) using drop-down menus at the bottom of the screen. All the necessary information to complete the game plan is provided on a single screen, and no training to use the system is required. Figure 7.1 shows an example scenario from ATTAC.

Once the trainee submits their game plan, ATTAC assesses the trainee on a three-point scale based on whether the game plan would be safe and effective to meet the commander's intent. As previously mentioned, each game plan had to be considered holistically, since the individual game plan TMOI components are dependent on one another and on various features of a given scenario. In addition, game plans could not be assessed simply as correct vs. incorrect decisions, since some game plans may meet the goal of the mission but contain some judgment error. As a result, each game plan could be scored as ideal, acceptable, or unacceptable, and we based this judgment on the likelihood of meeting commander's intent. Ideal game plans were the best possible

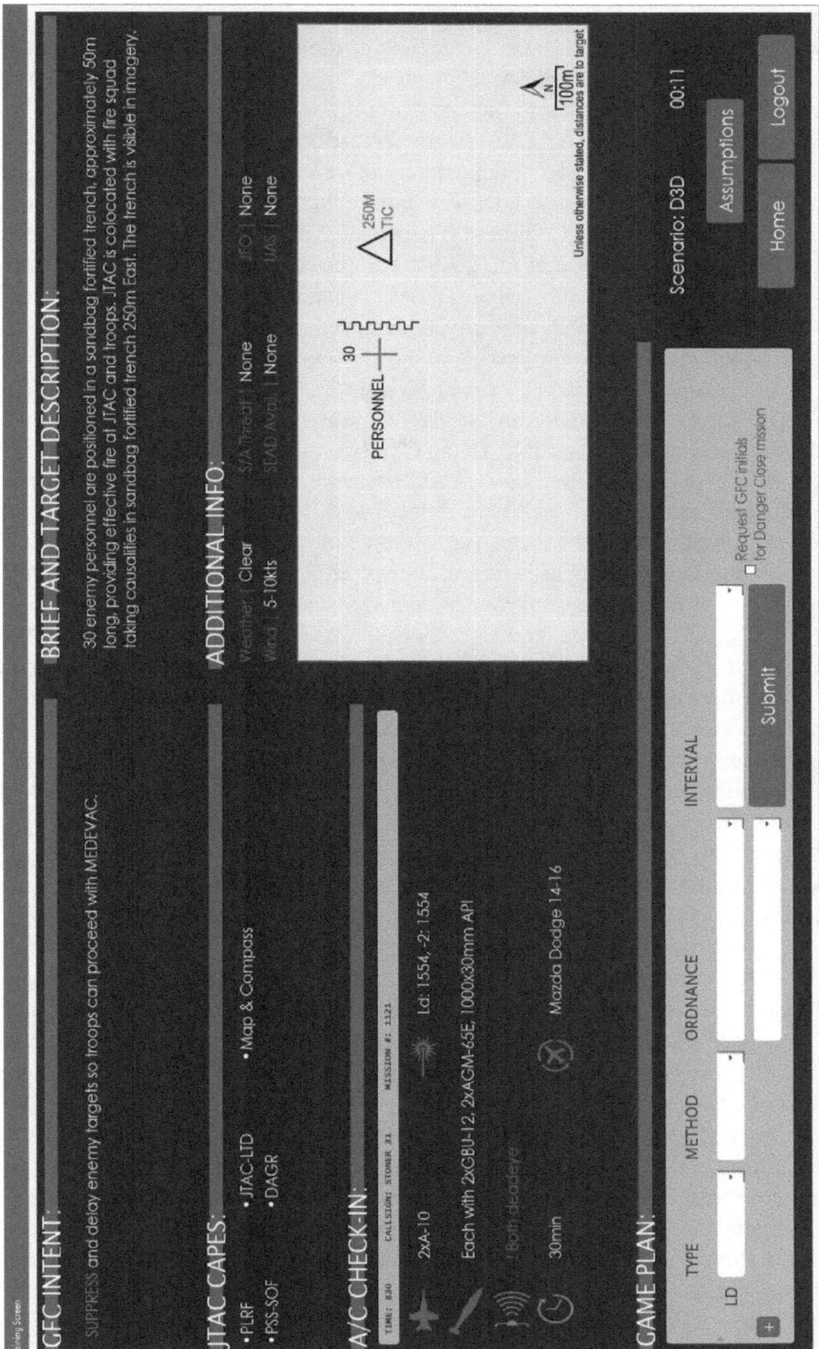

FIGURE 7.1 Example of an ATTAC scenario (Intermediate difficulty).

decisions that would meet commander's intent effectively and efficiently. Acceptable game plans may have met commander's intent, but they could have been improved in some way (e.g., it may not have been as efficient or the best weapon-to-target match). Finally, unacceptable game plans would likely not meet commander's intent or may not have been possible altogether in the context of the scenario.

Based on this assessment, ATTAC adaptively responds with two different instructional interventions, adaptive feedback and adaptive scenario difficulty. The underlying mechanisms for how these adaptive interventions worked were grounded in principles derived from the Cognitive Theory of Multimedia Learning (CTML; Mayer, 2021; see also Johnson et al., 2019 for a deeper discussion of the cognitive theory-driven approach to the design of ATTAC). In short, CTML states that learners have a limited working memory capacity, so instruction must be carefully designed to foster productive cognitive processing and limit unproductive processing to avoid situations of cognitive overload (Mayer & Moreno, 2003). First, ATTAC adapts the type of feedback the trainee receives based on the trainee's response to each scenario and points to doctrine when possible. Using CTML as a guide, we designed feedback messages of varying detail based on ATTAC's assessment of the trainee's game plan (see Table 7.1 for examples of each type of feedback). For ideal game plans, ATTAC provides positive outcome feedback (i.e., "Good job!"). Since the trainee's game plan was correct, any additional information would be extraneous and may impose unproductive cognitive processing on the learner (Kalyuga, 2007). For acceptable game plans, ATTAC provides feedback about the specific element that could have been improved in the game plan. Since the trainee's game plan was mostly correct, the feedback message is targeted to discuss only the element that needed improvement with the aim to focus the trainee on the necessary information and reduce any unproductive extraneous cognitive processing. Finally, for unacceptable game plans, ATTAC provides detailed feedback with a description of the thought process behind each game plan TMOI element. In this case, the detailed feedback is necessary to reduce extraneous cognitive processing and foster productive cognitive processing (Johnson & Marraffino, 2022; Moreno 2004), so that the learner receives information necessary to understand how an expert would approach the situation.

Second, based on the assessment of a series of two scenarios, ATTAC adjusts the difficulty of subsequent scenarios with the intention of managing the trainee's cognitive processing demands (see Wickens et al., 2013 for a review of adaptive difficulty). ATTAC includes scenarios of basic, intermediate, and advanced difficulty, which is determined by the complexity of the scenario (i.e., the number of interacting elements that one must consider when making a game plan decision). For example, when trainees are struggling and submitting unacceptable game plans, the next set of scenarios will be at an easier level of difficulty. Likewise, if trainees are performing well and submitting ideal game plans, then they will receive more difficult scenarios. Finally, for trainees who are performing in the middle, they maintain the same level of difficulty. Trainees are aware of the scenario's difficulty level as it is indicated by the color of the framing of the scenario screen (i.e., green is basic, yellow is intermediate, orange is advanced).

In summary, ATTAC was designed with a simple, easy-to-use interface to allow JTAC trainees to practice game plan development skills. We relied heavily on the CTML to drive the design of the adaptive feedback messages and adaptive scenario

TABLE 7.1
Examples of Feedback Messages for Ideal, Acceptable, and Unacceptable Game Plans

Ideal

Top Level Feedback: Good job! Your game plan should meet Commander's Intent…

	Aircraft 1			Aircraft 2			
	Type	**Method**	**Ordnance**	**Type**	**Method**	**Ordnance**	Interval
Your game plan	Type 1	BOC	30 mm API	Type 1	BOT	30 mm API	30 s–1 min

Acceptable

Top Level Feedback: Almost there. Your game plan would work, but consider the following adjustments:

	Aircraft 1			Aircraft 2			
	Type	**Method**	**Ordnance**	**Type**	**Method**	**Ordnance**	**Interval**
Consider another game plan	Type 1	BOC	30 mm API	Type 1	BOT	30 mm API	30 s–1 min
Your game plan	Type 1	BOC	30 mm API	Type 1	BOT	30 mm API	**1 min–2 min**

Specific Feedback: 30 s–1 min spacing is recommended for unguided munitions (Reference). Greater spacing may result in missed opportunities because it gives targets time to escape.

Unacceptable

Top Level Feedback: Not quite. Your game plan is unlikely to be successful.

	Aircraft 1			Aircraft 2			
	Type	**Method**	**Ordnance**	**Type**	**Method**	**Ordnance**	**Interval**
Consider another game plan	Type 1, 2, or 3	BOC	30 mm API	Type 1, 2, or 3	BOT	30 mm API	30 s–1 min
Your game plan	Type 1	BOC	GBU-12	Type 1	BOT	GBU-12	1 min–2 min

Detailed Feedback:

- Both A/C are deadeye. Ground-based lasing with other available ordnance is not recommended due to JTAC/TGO operator survivability. Therefore, guns are a recommended target-weapon match for suppressive effects.
- Type 1 or Type 2 is acceptable for GP ordnance based on JTAC tactical risk assessment.
- It would be acceptable for Ld to start with a BOC attack with guns for suppressive effects and provide a mark for –2 to follow on with corrections.
- If a Type 1 or 2 attack is utilized, 30 s–1 min is recommended for unguided munitions (Reference). Greater spacing may result in missed opportunities because it gives targets time to escape. A window of engagement would be appropriate for Type 3.

difficulty to manage an individual trainee's cognitive resources effectively. In the following section, we discuss two evaluations of ATTAC conducted with USMC students to determine whether using ATTAC helped students improve their game plan decision-making performance.

IS ATTAC EFFECTIVE?

Controlled Experiment

To assess the training effectiveness of ATTAC, we conducted a study with students enrolled in the Joint Fires Observers (JFOs) course. JFOs engage in similar tasks as JTACs, but JFOs do not have authority to grant weapons release during CAS missions. Moreover, JFO students complete much of the same coursework as JTAC students in the Tactical Air Control Party (TACP) course, which is a required course for JTAC certification. Therefore, JFO students were a suitable population for this experiment.

Details of the controlled experiment can be found in Marraffino and colleagues (2019), but to summarize here, 52 Marines participated in one of three training conditions. In the Adaptive condition, they completed 35 minutes of training with the adaptive version of ATTAC as described above. In the Non-adaptive condition, they completed 35 minutes of training with a version of ATTAC that kept the feedback and scenario difficulty constant, regardless of how students performed during training. In the Control condition, students reviewed slides about game plan development. Prior to training, all students completed a pre-test that comprised nine game plan scenarios of various difficulties and did not receive feedback on their performance; these test scenarios were not included in the library of ATTAC scenarios, so students were unable to train on them. Next, students completed their assigned training condition, and then they completed the post-test, which included the same items as the pre-test but in a different order. As shown in Figure 7.2, students in the Adaptive condition had the highest pre- to post-test increase compared to those in the Non-adaptive and Control conditions. In fact, when computing gain scores that accounted for how much students could have improved from pre- to post-test (i.e., difference between post- and pre-test scores divided by the difference between total possible score and pre-test), the results showed that the Adaptive condition's gain scores were 400% higher than those in the Control condition and 118% higher than those in the Non-adaptive condition. These results demonstrated that training on ATTAC for only 35 minutes produced measurable learning gains and that adaptive training was an effective technique for a complex decision-making task like planning CAS attacks.

Classroom-based Evaluation

Based on the success of the controlled experiment, our research team was invited to evaluate ATTAC in the context of a USMC course, specifically in the TACP Primer course. This course is typically offered to Marines in artillery units to prepare them for the TACP course, and it goes over some of the material that will be covered during the TACP course, including game plan development.

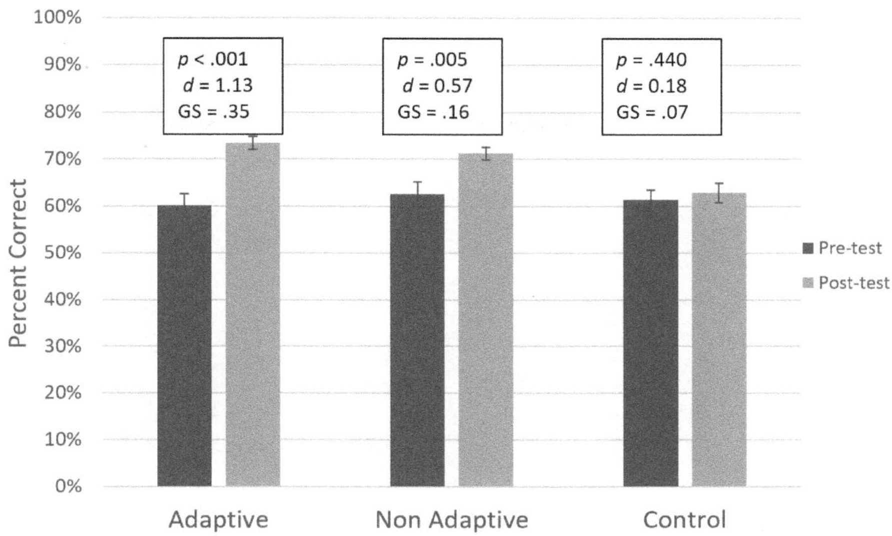

FIGURE 7.2 Controlled experiment pre-test and post-test performance by condition.

Note: p is p value. d is Cohen's d with $d = 0.8$ generally considered a large effect size, $d = 0.5$ a medium effect size, and $d = 0.2$ a small effect size (Cohen, 1988). GS refers to gain score: (post-test – pre-test)/(max score – pre-test).

Following the game plan lecture, seven Marines enrolled in the TACP Primer course completed the pre-test, spent 35 minutes training with the adaptive version of ATTAC, and then completed the post-test (i.e., the same procedure as the controlled experiment but with all participants assigned to the Adaptive condition). We scored the tests and found that six out of seven students improved from pre-test to post-test (one student did not follow the instructions and failed to fully answer each question on the post-test). The lead instructor remarked, "there's no 35 min lecture I could give to get learning gains like that." As a result, the following day, the instructors opted for students to continue to use ATTAC for over one hour during class time, while the instructors circulated around the room and interacted with the students. Since ATTAC saliently displays the difficulty of the scenarios, instructors could easily tell which students were having difficulty, because they remained on the basic difficulty scenarios, and which students were performing well, because they were receiving intermediate and advanced scenarios. The instructors used the context of the scenarios to ask the students questions about why they made certain decisions and to explain why some decisions would be beneficial in certain situations. At the end of this session, we again administered a post-test. This time, we found all seven students to improve relative to their pre-test score (see Figure 7.3). Although this was not a controlled experiment, the results revealed that students benefitted from getting reps and sets while using ATTAC on its own and also with instructor intervention. Taken together, both studies demonstrated ATTAC to be effective for helping

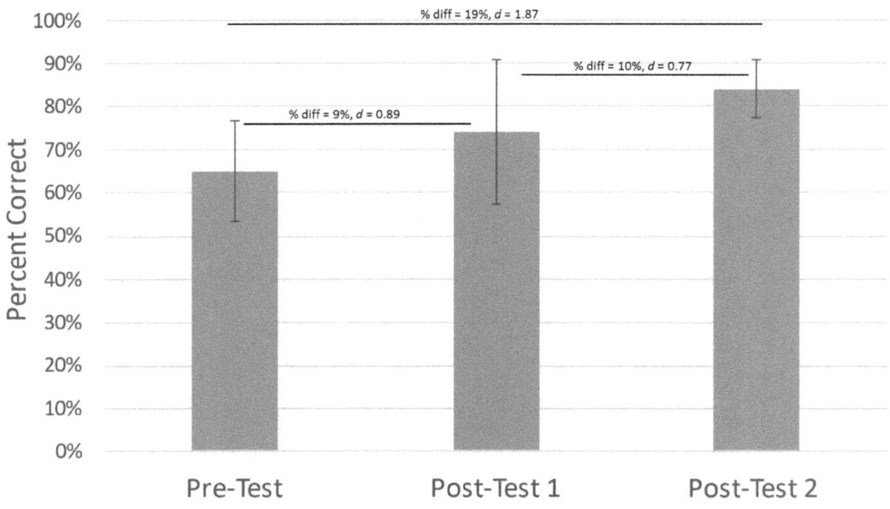

FIGURE 7.3 Classroom-based evaluation pre-test, post-test, and second post-test performance.

Note: diff is the difference between two test scores. d is Cohen's d with $d = 0.8$ generally considered a large effect size, $d = 0.5$ a medium effect size, and $d = 0.2$ a small effect size (Cohen, 1988).

students improve their game plan decision-making skills with and without the help of an instructor. Consequently, several units decided to continue using ATTAC in their courses to provide students practice with game plan development, and it is now officially included in a suite of USMC training tools.

WHAT WAS CHALLENGING DURING ATTAC DEVELOPMENT AND HOW COULD AI/ML HELP?

Given the complexity of the task and the extensive domain knowledge required to conduct it, scenario development, assessment, and feedback generation were challenges that had to be overcome in the process of building ATTAC. In the early stages of the effort, each of these tasks required a manual approach and significant input and iteration with subject matter experts (SMEs). Our success with this manual approach led us to continue this way over the course of the project and build over 100 scenarios and their associated assessments and feedback statements. But now with the benefit of hindsight (and a computer scientist on the team), we recognize that AI/ML approaches likely would have been beneficial once we had a corpus of validated scenarios to help us to build new scenarios. In the following sections, we step through the process of building ATTAC, describe these challenges in more detail, and offer areas of opportunity where AI/ML approaches (see Mahesh, 2020 for a review) could assist in making these processes more efficient.

Scenario Development Process

Creating scenarios presented two primary challenges. First, scenarios needed to be relevant to the decision-making process for game plan development and cover a sufficiently broad range of missions. To tackle relevancy, we first identified the critical information required for JTACs to make their game plan decisions by interviewing several TACP instructors and SMEs, observing lectures and scenario-based simulations, and reviewing course materials. The overall design of ATTAC was inspired by a homework assignment provided during the game plan portion of the TACP course. It included a series of short text-based vignettes comprising a commander's intent; the aircraft check-in; and the type, number, and location of the target and nearby friendly forces from which students derived their game plan. Further discussions with instructors and SMEs led to additional scenario elements that would help drive game plans and increase the complexity of the decision-making process (weather conditions, time on station, presence of anti-air threats in the area, etc.).

Once the key scenario elements were identified, the next hurdle was developing scenarios with sufficient realism. Instructors and SMEs emphasized the need for scenarios to be realistic to build JTAC trainees' experience and task fluency to prepare them for the types of missions that they may face during deployment. Critically, scenarios needed to include target sets that were relevant to the current priorities of the USMC, and targets needed to be placed in locations where a CAS mission could be successful (e.g., placing targets close enough to friendlies to warrant CAS but not so close that a mission could not be carried out safely). Some other scenario realism considerations to be made included ensuring that aircraft check-ins were believable, such that the attacking aircraft were carrying ordnance compatible with the platform (and in the right quantity), down to details such as having realistic call signs, mission numbers, and laser codes to get JTAC trainees oriented to the rhythm associated with a real aircraft check-in. Overall, all the scenario elements needed to be arranged carefully with an eye for realism to create a variety of scenarios that meaningfully captures the breadth of CAS missions that JTACs may experience in real-world operations.

The second challenge with scenario development was creating a scenario library that contained a variety of difficulty levels (i.e., advanced, intermediate, and basic) so that ATTAC could adapt scenario difficulty based on the trainee's performance. From a CTML perspective, scenario difficulty was based on the number of interacting elements within each scenario that need to be handled in working memory during the decision-making process. To that end, working with instructors and SMEs, we identified areas within the scenario elements that could increase (or decrease) the number of considerations a trainee would have to draw upon to make an informed game plan. One example was to edit the number of weapons available to carry out the mission. Having fewer options from which to choose reduces the difficulty of the ordnance decision, whereas having more options increases the difficulty. Other examples of increasing scenario difficulty include adding anti-air threats to the scenario, including weather complications (e.g., dense cloud layers), and placing friendlies in closer positions to the target, because all these considerations may have an impact on the game plan decisions.

To develop scenarios with varying difficulty, we first created a sample set of scenarios targeted at an intermediate difficulty. From this set of base scenarios, we created advanced and basic versions by tweaking the elements described previously. For example, to create a basic scenario, we removed elements, and to create advanced scenarios, we added elements. To validate these initial scenarios, several instructors and SMEs provided feedback on the relevance and realism and rated each scenario's difficulty. The result was a set of scenarios that we could use as templates to expand ATTAC's training library.

Reflecting on the processes for scenario creation, we recognize that AI/ML approaches could have been helpful to build scenarios with relevance to USMC training and assign an appropriate difficulty level to each one. The combination of SME input regarding critical scenario elements, SME difficulty assessments, and actual trainee performance metrics on these scenarios that we aggregated during ATTAC development can serve as labeled data sets for building models of scenario relevance and difficulty. This approach could allow for partially or fully automated scenario creation and difficulty classification in the future. For example, given the breadth of critical elements featured in our set of scenarios created with SME input, we could build supervised learning models that predict which combinations of such elements are realistic for game plan training scenarios. Additionally, we could develop classifiers that predict the difficulty of a scenario or clustering algorithms that group scenario difficulty given the presence or absence of these elements. Where available, actual trainee performance metrics can also serve as an input to these models. These two models together could assist in the creation of new scenarios, whether by classifying the relevance and difficulty of human-generated scenarios or by developing generative models that themselves create new scenarios.

Performance Assessment Process

Assessment was another challenging and manual process to overcome during ATTAC's development. During initial conversations with instructors and SMEs, it became clear that each individual game plan component (i.e., TMOI) could not be scored in isolation, and it was most often the case that multiple game plans could successfully accomplish the mission. This posed two challenges. First, because game plan components could not be scored individually, the number of game plan combinations to consider for each scenario increased exponentially. Second, the assessment needed to be able to account for multiple correct answers. This created a situation that for any given scenario, there could be upward of 1,600 possible game plan combinations that would need to be evaluated, with multiple game plans that could be scored as ideal or acceptable. To address these issues, we provided instructors and SMEs a sample of scenarios we created and asked them to identify ideal game plans (i.e., game plans that would efficiently satisfy the commander's intent) and acceptable game plans (i.e., game plans that may meet mission requirements but may not have been the most efficient or effective approach) for each. We also followed up with interviews to discuss discrepancies across SMEs. Based on these discussions, we looked for patterns in their game plan selections to create a series of heuristics to aid in identifying ideal and acceptable answers for other scenarios.

Using our heuristics, we first identified all ideal game plans for a scenario, then we tested additional game plan combinations that might be considered acceptable, which generally concerned game plans that would likely take more time to execute or were not the best weapon-to-target match when compared to an ideal game plan. Once we selected ideal and acceptable game plans for each scenario, all combinations that were not specifically identified in the initial evaluation were considered unacceptable. This process was repeated for each of the approximately 100 scenarios created for the ATTAC library, resulting in a database of over 1,000 entries of ideal and acceptable game plans. Afterward, instructors and SMEs validated these for each of the scenarios. To drive assessment within ATTAC, the trainee's game plan was compared to all possible ideal and acceptable game plans in the database for that scenario. If a match was found, the trainee was assessed accordingly. If no matches were found, the trainee's game plan was scored as unacceptable.

As with scenario creation, game plan assessment could potentially benefit from AI/ML approaches. Our collection of assessed game plans across our ATTAC scenario library can serve as labeled training data for building a game plan assessment classifier. As input, such classifiers would use the major components of a game plan (i.e., TMOI). Additionally, they would need to consider other specific aspects of each scenario, such as the location of friendlies and weather conditions, as these must be factored into game plan selection and considered during assessment. As output, these classifiers would assign a game plan assessment of ideal, acceptable, or unacceptable. The creation of such classifiers might require a larger database of scenarios that collectively encompasses the domain of possible combinations of inputs to ensure that they can effectively model game plan assessment.

By building classifiers over these labeled training data, we can automate the classification of game plans on new scenarios; as opposed to manually deriving patterns from SME-driven assessments of ideal and acceptable game plans to assess the remaining possible game plans, we could simply classify all game plans available for a given scenario. This would replace the current database of game plan assessments and simplify future maintenance of ATTAC.

Developing Feedback Statements

For each assessment, we generated feedback statements for acceptable and unacceptable game plans (ideal game plans only received minimal outcome feedback that said, "Good job!"). For unacceptable game plans, trainees received a detailed description of the reasoning behind an alternative ideal game plan broken out by TMOI decisions. Detailed feedback messages were generated with instructors and SMEs for the sample scenarios, which were used as templates to manually generate additional feedback statements for other scenarios. Each series of feedback statements considered the holistic nature of the game plan to discuss how a given game plan element (e.g., TMOI) worked with the rest of the game plan and included references to doctrine when available. However, for many scenarios, multiple game plans could be considered ideal. To reduce confusion in these cases, we developed an algorithm that selected the closest possible match to the trainee's submitted game plan. In this way, we could tailor feedback to the trainee that was the closest to their way of thinking. Although this approach helped align the feedback with the trainee's intent, it also meant generating additional sets of feedback statements for every possible ideal game plan for each scenario. Overall, since there were over 1,000 ideal and acceptable game plans included in our database, our research team generated an equivalent number of feedback statements.

For acceptable game plans, the team identified the element(s) of the game plan during the assessment process that could be modified to nudge the submitted game plan into one that was ideal. For instance, if the submitted interval needed to be shorter, the feedback statement only discussed the interval decision and how it could be improved within the context of scenario. Any given scenario could have dozens of acceptable game plans and their associated feedback statements.

One avenue for automating feedback statement generation is the development of a domain-specific language model. While our feedback statements have some natural variance in sentence structure and word choice, the overall domain of possible statements and topics is fairly limited, as these statements focus on explaining the advantages and disadvantages of a finite set of game plan components and other scenario attributes. Many of these statements may apply to a large number of scenarios, such as those with specific weather conditions that limit the effectiveness of a particular weapon. As such, a language model trained on these scenario components and on doctrine could drive the formation of realistic, useful feedback statements to present to trainees based on their specific game plan selections.

LOOKING AHEAD, HOW COULD AI/ML IMPROVE ATTAC's RESPONSE ALGORITHMS?

Although ATTAC is currently a fielded system, there may be opportunities to improve it over time as more data are collected from trainees. A particular strength afforded by adaptive training is the ability to dynamically update instruction in response to the individual needs of each learner, such as by adjusting difficulty or presenting additional training examples for topics that the learner consistently misses. Ideally, an adaptive trainer would target a level of difficulty that promotes skill acquisition for a given learner, such that training content is neither too hard nor too easy. Given our collection of actual trainee performance data, there is potential to incorporate AI/ML algorithms to fine-tune ATTAC's adaptive capabilities to better deliver an appropriate level of challenge to each trainee, which we predict would lead to more efficient and effective training.

Currently, ATTAC decides to increase, decrease, or maintain scenario difficulty based on the number of points awarded to the trainee's game plans across consecutive scenarios, with ideal game plans receiving the most points. As such, ATTAC's ability to measure skill acquisition is facilitated solely by a holistic view of learner game plan assessment over time without consideration of the patterns of incorrect responses or aspects of the particular scenario. In other words, ATTAC's assessment algorithms could be improved by including a more sophisticated diagnosis of the types of scenarios with which a trainee is struggling. For example, it could be the case that a trainee struggles with choosing the correct Interval, given a particular

Ordnance. Likewise, scenarios across different difficulty levels may share specific scenario elements that a trainee is less comfortable with, such as how weather conditions affect Method decisions or their familiarity with the capabilities of the attacking aircraft. This type of approach could expand how difficulty is structured within ATTAC to deliver targeted scenario elements and better quantify specific skill acquisition beyond just overall game plan assessment.

For a given trainee, ATTAC could continuously build an individualized model that predicts the performance assessment for scenarios with given attributes. Such models may better represent clusters of scenarios on which a given trainee would perform well or poorly than only scenario difficulty. Rather than simply adapting difficulty up or down based on consecutive game plan assessments, ATTAC could queue up scenarios on which the trainee is expected to receive acceptable assessments to target an appropriate difficulty level. Additionally, for clusters in which a trainee is likely to perform poorly, ATTAC could begin by delivering easier scenarios to build up the trainee's familiarity with this specific type of scenario. Finally, ATTAC could use the average predicted assessment of each cluster as means of determining when training should end.

HOW CAN AI/ML BE USED TO MAINTAIN ATTAC?

One major challenge associated with developing training solutions like ATTAC is the ongoing task of adapting to changes in military technologies, capabilities, tactics, techniques, and procedures. The dynamic nature of military practices creates a requirement for frequent maintenance of most training systems to ensure that the training content is fresh and relevant. CAS training needs to evolve quickly as new weapons capabilities are introduced (while older systems are phased out); tactics, techniques, and procedures change in response to lessons learned and new capabilities; and areas of operation are updated as new threats emerge. Therefore, it is important not only to create effective training scenarios at the start but also to ensure that they remain relevant and aligned with ever-changing military strategies and technologies throughout the lifecycle of the training system.

During ATTAC's initial three-year development cycle, we encountered two major changes that highlighted the need to update the scenario library to maintain currency for training. First, after establishing the initial set of ATTAC scenarios, new doctrinal publications were released that significantly altered the deployment strategies for certain types of weapons. These modifications directly affected how ATTAC should assess game plans and what the feedback statements should say for numerous scenarios that included these weapons, necessitating an immediate update to the database. This process required careful attention to detail to make sure that ATTAC accurately reflected the latest doctrine, which demanded not only an in-depth understanding of the changes but also a meticulous updating process to ensure accuracy and consistency.

Second, two new weapon systems became available for CAS missions, which necessitated additional modifications to the scenario library. Instructors not only requested the inclusion of these new weapons in existing scenarios to replace outdated systems, but also new scenarios specifically designed around these capabilities.

Once again, this required manual updates to scenarios and ensured that the associated assessments and feedback aligned with the proper utilization of these new weapon systems. These challenges emphasized the need to efficiently integrate changes in both doctrine and emerging technologies to ensure that training remains effective and aligned with the dynamic nature of military practices.

AI/ML techniques could be applied to improve the processes necessary to maintain the library of scenarios and their associated assessments and feedback statements. Major manual changes to the databases pose the risk of unintended errors, such as reducing the accuracy and relevance of scenario content and the consistency of assessments, especially as military practices and training needs change over time. However, given updated SME input, the aforementioned AI/ML models for scenario relevance and difficulty, game plan assessments, and natural language feedback statements can themselves be updated and rebuilt in response to changing needs.

In general, this would manifest as updates to the labels and weights of the data used to build these models. For example, as USMC priorities change, specific combinations of scenario components may lose relevance or realism, so these components can be de-emphasized or otherwise removed from consideration when modeling. For natural language models, this would involve changes to the input domain of words and topics. In other words, we remap the problem from manually updating database entries to tweaking model parameters, which can be smaller in scope, more reliable, and more generalizable. Manual database updates may be subject to human error, such as inconsistently applying relevant changes across the database, and these risks would be present every time training content requires modification. However, any time models are rebuilt, all resulting updates are directly and automatically applied to the entire database of scenarios, game plans, and feedback assessments. As such, the same overall framework for building models based on SME input is still applicable, reducing the cost of maintenance once these processes are established. Given the crucial nature of domain expert input, we could provide summaries of these changes for subsets of affected training content, such as scenarios featuring newly available weapons, to SMEs for validation, ensuring that any model updates produce accurate, consistent results.

HOW CAN WE IMPROVE INSTRUCTOR CONFIDENCE IN AI/ML APPROACHES?

In our experience, some instructors are hesitant to adopt adaptive training technologies, AI/ML notwithstanding. While AI/ML can provide numerous benefits for adaptive training system development, there are some notable concerns that might affect instructor adoption. In particular, we will focus on issues such as lack of confidence or trust, job security, unrealistic expectations, and safety concerns (Cubric, 2020) and discuss how they relate to incorporating these capabilities into ATTAC to enhance scenario generation, game plan assessment, feedback, and adaptive training capabilities. Of these focal areas, improving adaptivity through AI/ML is likely to be readily accepted, especially given existing studies already showing improved training outcomes through adaptive approaches, particularly within military tasks

(Barto et al., 2020; Bond et al., 2019; Landsberg, Mercado et al., 2012; Marraffino et al., 2019; Van Buskirk et al., 2019; Whitmer et al., 2021). However, instructors may be hesitant to trust training content built upon AI or ML technologies, since they understand the inherent complexities of generating relevant scenarios at prescribed difficulty levels and providing appropriate assessments of and feedback for game plans.

Ultimately, we propose two main avenues to satisfy these reasonable concerns. Most importantly, we found that establishing continuous input and validation from SMEs was vital to developing ATTAC in the first place, and this would remain true even when moving to more advanced AI/ML techniques. ML models are a potential tool for enhancing and better incorporating this domain-specific knowledge within ATTAC, not for replacing it or the need for expert instructors. Our current data sets of scenario difficulties, game plan assessments, feedback statements, and learner performance can all serve as ground truth data for ML algorithms to model. Providing instructors with measurements of how closely these models match this ground truth could promote confidence in their reliability and correctness. Additionally, prior to any use in the classroom, we would intend to have instructors fully validate any content generated by these AI/ML enhancements, essentially continuing the validation steps we performed with SMEs during initial ATTAC development. Though less common than other AI validation methods (Myllyaho et al., 2021), domain expert validation is vital to ensuring the correctness and consistency of new or updated scenario content due to the nuanced nature of game plan development and assessment, and continued involvement with instructors will likely assuage concerns of scenario accuracy and consistency and result in more acceptance of these techniques. Validation of automated game plan assessments is especially important, as these selections have real-world safety concerns.

Instructors familiar with large language models (LLMs), such as ChatGPT, may have concerns about the quality and correctness of dynamically generated feedback statements and about the capabilities and reliability of language models in general (Kasneci et al., 2023). These may also be more difficult to validate with instructors, given that each trainee may receive entirely customized feedback. As such, we suggest significant SME involvement in building and testing the language model to ensure that it employs appropriate words and sentence structures, along with continued monitoring during training. Furthermore, it is important for instructors to understand that a language model for a task like ATTAC feedback generation has a significantly restricted domain compared to an LLM, which may help ease quality concerns.

Finally, it is important to reiterate that the end product of the training will provide direct benefits to instructors, because students may gain individualized reps and sets when instructors are unavailable. For example, this can enable all students to come into a classroom with a baseline knowledge and understanding, allowing instructors to teach to a common baseline and provide opportunities to discuss more advanced topics when there may otherwise not be time. In addition, it can also help students avoid skill decay because they can use the training to practice the skills they learned periodically during the course and potentially as refresher training well after the course has ended.

DISCUSSION

In this chapter, we presented ATTAC as a use case to describe the manual processes we undertook to develop an adaptive scenario-based trainer for a complex military decision-making task and the challenges with maintaining it. After reflecting on these challenges, we proposed AI/ML approaches that could increase the efficiency of some of these processes and facilitate development and maintenance. For scenario creation and assessment, we highlighted supervised learning and classification approaches that could leverage the existing work product to facilitate rapid creation of additional content. For feedback generation, we described how a domain-specific LLM could reduce the workload associated with creating realistic and useful feedback statements. With these AI/ML approaches, we could not only efficiently add to the existing scenario library to increase the variety of scenarios delivered to students but also reduce the workload for maintenance when new weapons, capabilities, and equipment emerge and tactics, techniques, and procedures evolve over time. Incorporating these approaches earlier in our process may have also increased the efficiency of initial content creation. As a future research direction, we noted how clustering approaches could be employed to improve the adaptive capability of ATTAC by providing scenarios designed to target more specific knowledge gaps and situations where additional practice would maximize learning gains. Importantly, these approaches open the door to conduct research determining best-practices and identifying novel ways to employ adaptive training that maximize learning outcomes in complex domains. AI/ML approaches provide opportunities to better assess behaviors and their contributing factors and provide more tailored instruction.

Lastly, a lack of acceptance of AI/ML-assistive technology may be a barrier to its adoption. Therefore, it is critical that developers involve instructors in discussions about the design and application of the algorithms and their role in the system. Instructors should also be brought in to verify and validate some or all of the AI-generated data. Moreover, it is important to ensure that instructors understand that the AI/ML algorithms are a supplement to development. The expertise still lies with the instructors and the goal of these approaches is not to create expertise but to capture it.

CONCLUSION

As U.S. military training and education modernization efforts continue to expand across the services, adaptive training approaches are likely to play a growing role in their training pipelines, since they have been demonstrated to increase learning outcomes without placing additional burdens on instructors' time. Historically, building effective adaptive training systems has been a relatively costly and time-consuming endeavor, limiting access to this technology to particular use cases. But recent advances in AI/ML have the potential to speed up development and reduce costs, opening the door to potential widespread adoption of adaptive training solutions for many more applications.

ACRONYMS AND ABBREVIATIONS

AI	Artificial Intelligence
ATTAC	Adaptive Trainer for Terminal Attack Controllers
BOC	Bomb on Coordinate
BOT	Bomb on Target
CAS	Close Air Support
CTML	Cognitive Theory of Multimedia Learning
DoD	Department of Defense
JFO	Joint Fires Observer
JTAC	Joint Terminal Attack Controller
LLM	Large Language Model
ML	Machine Learning
NLP	Natural Language Processing
SME	Subject Matter Expert
TACP	Tactical Air Control Party
TMOI	Type, Method, Ordnance, and Interval
USMC	United States Marine Corps

ACKNOWLEDGMENTS

This work was funded by Dr. Peter Squire at the Office of Naval Research (funding document # N0001424WX00190). We gratefully acknowledge Dr. Daphne Whitmer who was instrumental in developing the testbed and participating in the data collection efforts described herein.

DISCLAIMER

Presentation of this material does not constitute or imply its endorsement, recommendation, or favoring by the U.S. Navy or Department of Defense (DoD). The opinions of the authors expressed herein do not necessarily state or reflect those of the U.S. Navy or DoD.

REFERENCES

Barto, J., Daly, T., LaFleur, A., & Steinhauser, N. B. (2020, December). Blending adaptive learning into military formal school courses. *Proceedings of the Interservice/Industry Training, Simulation, and Education Conference*. Orlando, FL: National Training Systems Association.

Berger, D. H. (2019). 38th Commandant's Planning Guidance (SSIC No. 05000 General Admin & Management).

Bond, A. J. H., Phillips, J. K., Steinhauser, N. B., & Stensrud, B. (2019, December). Revolutionizing formal school learning with adaptive training. *Proceedings of the Interservice/Industry Training, Simulation, and Education Conference*. Orlando, FL: National Training Systems Association.

Cohen, J. (1988). *Statistical Power Analysis for the Behavioral Sciences* (2nd ed.) Hillsdale, NJ: Erlbaum.

Cubric, M. (2020). Drivers, barriers and social considerations for AI adoption in business and management: A tertiary study. *Technology in Society*, *62*, 101257.

Gilday, M. (2022). Chief of Naval Operations Navigation Plan. https://www.dvidshub.net/publication/issues/64582

Johnson, C. I., & Marraffino, M. D. (2022). The Feedback Principle in Multimedia learning. In R. E. Mayer, & L. Fiorella (Eds.), *The Cambridge Handbook of Multimedia learning*, 3rd ed. (pp. 403–417). Cambridge, UK: Cambridge University Press.

Johnson, C. I., Marraffino, M. D., Whitmer, D. W., & Bailey, S. K. T. (2019). Developing an adaptive trainer for joint terminal attack controllers. In R. A. Sottilare, & J. Schwarz (Eds.), *HCII 2019: Adaptive Instructional Systems. Lecture Notes in Computer Science 11597* (pp. 314–326). Cham: Springer.

Kalyuga, S. (2007). Expertise reversal effect and its implications for learner-tailored instruction. *Educational Psychology Review*, *19*, 509–539.

Kasneci, E., Seßler, K., Küchemann, S., Bannert, M., Dementieva, D., Fischer, F., …, & Kasneci, G. (2023). ChatGPT for good? On opportunities and challenges of large language models for education. *Learning and Individual Differences*, *103*, 102274.

Landsberg, C. R., Astwood, R. S., Van Buskirk, W. L., Townsend, L. N., Steinhauser, N. B., & Mercado, A. D. (2012). Review of adaptive training system techniques. *Military Psychology*, *24*(2), 96–113.

Landsberg, C. R., Mercado, A. D., Van Buskirk, W. L., Lineberry, M., & Steinhauser, N. (2012, September). Evaluation of an adaptive training system for submarine periscope operations. *Proceedings of the Human Factors and Ergonomics Society Annual Meeting*, *56*(1), 2422–2426.

Mahesh, B. (2020). Machine learning algorithms - a review. *International Journal of Science and Research*, *9*(1), 381–386.

Marraffino, M. D., Johnson, C. I., Whitmer, D. E., & Steinhauser, N. B. (2019). Advise when ready for game plan: Adaptive training for JTACs. *Proceedings of the Interservice/Industry Training, Simulation & Education Conference*. Orlando, FL: National Training & Simulation Association.

Mayer, R. E. (2021). *Multimedia learning* (3rd ed.). Cambridge, UK: Cambridge University Press.

Mayer, R. E., & Moreno, R. (2003). Nine ways to reduce cognitive load in Multimedia learning. *Educational Psychologist*, *38*, 43–52.

Moreno, R. (2004). Decreasing cognitive load for novice students: Effects of explanatory versus corrective feedback in discovery-based multimedia. *Instructional Science*, *32*, 99–113. https://doi.org/10.1023/B:TRUC.0000021811.66966.1d

Myllyaho, L., Raatikainen, M., Männistö, T., Mikkonen, T., & Nurminen, J. K. (2021). Systematic literature review of validation methods for AI systems. *Journal of Systems and Software*, *181*, 111050.

Park, O., & Lee, J. (2004). Adaptive Instructional Systems. In Jonassen, D.H. (Ed.), *Handbook of Research for Educational Communications and Technology* (2nd ed.) (pp. 651–684). Mahwah, NJ: Lawrence Erlbaum.

Shute, V. J., & Zapata-Rivera, D. (2012). Adaptive educational systems. *Adaptive Technologies for Training and Education*, *7*(27), 1–35.

U.S. Army Training and Doctrine Command. (2017). *Army Learning Concept for Training and Education for 2020–2040* (TRADOC Pamphlet 525-8-2). Ft. Eustis.

Van Buskirk, W. L., Fraulini, N. W., Schroeder, B. L., Johnson, C. I., & Marraffino, M. D. (2019). Application of Theory to the Development of an Adaptive System for a Submarine Electronic Warfare Task. In R. A. Sottilare, & J. Schwarz (Eds.), *HCII 2019: Adaptive Instructional Systems. Lecture Notes in Computer Science 1159* (pp. 353–362). Cham: Springer.

Whitmer, D. E., Johnson, C. I., Marraffino, M. D., & Hovorka, J. (2021). Using Adaptive Flashcards for Automotive Maintenance Training in the Wild. In R. A. Sottilare, & J. Schwarz (Eds.), *HCII 2021: Adaptive Instructional Systems. Lecture Notes in Computer Science 12792* (pp. 446–480). Cham: Springer.

Wickens, C. D., Hutchins, S., Carolan, T., & Cumming, J. (2013). Effectiveness of part-task training and increasing-difficulty training strategies: A meta-analysis approach. *Human Factors*, *55*(2), 461–470.

8 Exploring Cognitive Science Foundations for AI-Driven Healthcare Simulation

Shannon K. T. Bailey, Cheryl I. Johnson, and John Licato

INTRODUCTION

Healthcare professionals must be able to integrate vast amounts of knowledge to perform complex tasks with precision and efficiency, with a low margin for error. Medical errors are dangerous and costly, accounting for an estimated 40,000–90,000 deaths and costing up to $20 billion a year in the U.S. (Rodziewicz et al., 2023). Medical errors can be drastically reduced with deliberate practice, and simulation training is used to provide practice and evaluation of skills without risk to patients (Lioce et al., 2020). While we know simulation training can reduce medical errors (Cook et al., 2011; Okuda et al., 2009), training opportunities are often constrained by both time and resources as simulation typically involves instruction, practice, and evaluation facilitated by human subject matter experts. That demand on resources is only increasing as more training is needed to combat a shrinking healthcare workforce and keep up with an ever-expanding amount of medical knowledge. Because simulation is time and resource intensive, healthcare simulation is often presented with a "one size fits all" approach.

This "one size fits all" approach to simulation training is not always effective as individuals may differ in numerous ways. Cognitive psychology provides useful frameworks on how people think and learn that can be utilized to tailor healthcare simulation to individuals for better learning outcomes. The challenge of adapting healthcare simulation to an individual lies in scaling training that relies heavily on experts' time. Recent advances in artificial intelligence (AI) may help to make adaptive training feasible in healthcare simulation by reducing the amount of time experts needed during training, thereby expanding the capacity to train more learners. This chapter explores how theories from cognitive science can be combined with evolutions in AI to advance healthcare simulation, leading to more prepared medical professionals with fewer demands on an overburdened healthcare system.

 DOI: 10.1201/9781032701639-8

HEALTHCARE SIMULATION

Simulation for training healthcare professionals is utilized extensively across all specialties and levels of care, from teaching foundational tasks in undergraduate medical education to refreshing high-risk, low-frequency skills of experienced healthcare professionals. Simulation is used to teach a range of skills, from the steps to complete a medical procedure and diagnostic decision-making to effective communication within teams of healthcare professionals. To teach these diverse skills, healthcare simulation employs a vast array of simulation types, called *modalities* (Lioce et al., 2020). These simulation modalities may include human or animal elements, such as standardized patient actors (i.e., role players), animal models, and cadavers, though many healthcare training methods today involve use of technology. Healthcare simulation technologies range from physical simulators, including full-body manikins with varying degrees of anatomical fidelity and physiological responsiveness, to digital simulations, such as computer-based virtual patients or training scenarios in immersive environments (i.e., virtual reality [VR], augmented reality [AR], extended reality [XR]).

Choosing the appropriate simulation modality for training a certain skill is an important aspect to achieving desired learning outcomes. For example, if the learning objective is to teach students how to manage a medical emergency, the simulation should include key decision-making components, like requiring students to order appropriate medications, as well as communication within a team to help develop a mental schema of communication patterns. Different simulation modalities may be combined to achieve these learning goals.

What these simulation modalities and methods have in common is that they are often facilitated by subject matter experts, including clinicians, simulation operations specialists, and/or educators. Each training simulation usually requires developing the content, conducting the simulation, and evaluating trainee performance, all with a human expert or multiple experts. Continuing the above example, in a complex team scenario, it is difficult for a single expert to attend to every detail of the simulation, evaluate each trainee, and provide immediate, individualized feedback. If multiple evaluators are utilized, there may be differences among evaluators that lead to issues of inter-rater reliability. Developing and conducting simulation training is a time-intensive process for the expert, but also limits the number of trainees that can participate and the amount of individual tailoring that can be done during training, so there remains a lot of room for optimization in healthcare simulation.

Challenges in Simulation Training

There are many challenges to the current methods of healthcare simulation training, including the development of appropriate simulation content or scenarios, the accurate and timely assessment of performance, and the need to adapt the simulation to the learner's strengths and weaknesses to prioritize time spent in training. We provide an overview of these challenges in Table 8.1. While each simulation would ideally be specific to the task and trainees, this is often not practical with current

TABLE 8.1
List of Challenges in Healthcare Simulation Training

Challenge	Description	Possible AI Solutions
1. Providing timely, accurate assessment and feedback	Instructors typically evaluate trainees and provide feedback immediately following a simulation scenario, which has challenges, including inter-rater reliability, inconsistent feedback, or missing performance data.	AI may be used to adapt training to the individual trainee. Tailoring training to the individual's knowledge and performance is often more efficient and effective than "one size fits all" training methods.
2. Time to develop scenarios	Development of simulation scenarios typically involves determining the learning objectives, preparing simulation materials such as descriptions of symptom presentation or history of patients, determining what simulators and equipment will be required and how they will be implemented and by whom, defining assessment criteria, outlining pre-briefing and debriefing content, determining how the case will progress depending on learner performance, among many time-consuming considerations. Often, clinical educators must prepare their own simulation scenarios or simulationists must consult with busy clinicians to develop scenarios, and this work must be accomplished during limited time outside of patient care.	AI may be used to streamline the process of simulation scenario development by generating text-based planning of scenarios, including defining learning content, assessments, and logistics, saving time on clinician educators and simulation faculty. Generative AI may be useful in creating training scenarios, including the scenario text-based content, images of the clinical setting and patient, audio of the clinical encounter (e.g., patient speech and background noise), and sense of touch (i.e., haptic feedback).
3. Time during simulation	Instructors and trainees have restricted hours available to adequately train necessary knowledge and skills, so time in simulation must be optimized taking into consideration individual differences of learners (e.g., prior experience on the task) and the context for training (e.g., how many learners can participate at a time, is the training in-situ such that clinical resources are at a premium).	AI has the potential to offer scalable and reusable training opportunities with less space, staff, and resources. AI models that track trainee performance and provide tailored feedback could further increase the scalability and portability of training if scheduling around experts' availability is not required.
4. Training in low-resource environments	Clinical educators and high-fidelity simulations are not always located where training is needed, so either trainees or educators shoulder the burden of travel time and cost to on-site training. Distance simulation has been used to address geographical barriers to training, but these often have the same limitations as noted above related to time and resources.	AI may be used in distance simulation to assess performance and provide feedback, which may be useful in situations where an expert is not co-located with trainees.

simulation training methods; yet, advances in AI may be able to streamline these challenges in simulation training.

Furthermore, these challenges in developing and conducting simulation training are often interdependent. For example, developing simulation scenarios requires the educator to have identified the learning objectives and plan for how trainees will be assessed and debriefed. Performance assessment and feedback should be specific and timely, though this is not always the case if evaluators are limited in time or resources. We focus on the first of these challenges, *providing timely, accurate assessment and feedback*, as assessing learner understanding and debriefing is an integral part of effective simulation.

COGNITIVE FRAMEWORKS AND INDIVIDUAL DIFFERENCES

To tackle the challenge of delivering timely and accurate assessment and feedback during simulation, we can leverage frameworks from cognitive psychology. Drawing on decades of evidence identifying effective instructional strategies, we highlight approaches to enhance the delivery of adaptive simulation. When designing medical simulation training, one needs to consider the limitations of our cognitive architecture for it to be effective. The *Cognitive Theory of Multimedia Learning* or CTML (Mayer, 2020) describes how people learn from words and images and it provides a useful framework to consider when designing medical simulations. The central assumption of CTML is that learners' working memory capacity is limited, so it follows that instruction must be carefully designed to avoid overwhelming the learner's available cognitive resources. While people are learning, they actively engage in several cognitive processes. These processes include selecting the relevant words and images, organizing these words in a coherent verbal model and the images into a coherent pictorial model, and then integrating these models with each other and with their prior knowledge.

While they are engaging in these processes, there are three sources of demand on learners' cognitive processing resources. *Extraneous processing* stems from poorly designed instruction, such as including distracting information or a hard-to-use interface, in which learners engage in unproductive cognitive processing that is not relevant to their educational goal. *Essential processing* stems from the complexity of the material to-be-learned and is the cognitive processing necessary to mentally represent the concepts in their working memory. The learner's experience level can greatly affect essential processing, because as individuals are more experienced with material, they can chunk concepts efficiently; therefore, they can hold more information in their working memory than less experienced individuals. Finally, *generative processing* stems from the learner's effort to make sense of the information they are learning and is productive cognitive processing. These three processing demands are considered to be additive, so an increase in one leads to a decrease in capacity for the other two. Once a learner's cognitive processing resources have been depleted, it can result in learning decrements due to cognitive overload. Therefore, when designing instruction and simulations, one should aim to minimize extraneous processing, manage essential processing, and foster generative processing.

Cognitive Frameworks for Adaptive Training

Of course, we understand that individual learners are unique, and they come into a learning environment with different experiences and abilities, which play a significant role in how instructional design can affect their cognitive processing. One well-cited phenomenon that illustrates this point particularly well is the *expertise reversal effect* (ERE; Kalyuga, 2007, 2022). The ERE states that instructional techniques that may be beneficial for less knowledgeable learners may not be as effective (or can even be a detriment) for more knowledgeable learners. In a classic study by Kalyuga et al. (2001), inexperienced apprentices received training on programming relay circuits either by problem-solving (i.e., attempting to solve the problems themselves) or by worked examples (i.e., providing a similar problem with identical steps learners could apply to the problem along). The results showed that the worked examples group had higher learning outcomes than the problem-solving group. But over time as the apprentices became more knowledgeable about the domain as they trained, the benefits for worked examples washed away and the problem-solving technique became the more effective instructional strategy. These results suggest that the worked examples were initially helpful to inexperienced learners because they provided needed information to manage learners' essential processing. But as they developed expertise, the worked examples became redundant and distracting, which increased their extraneous processing demands and led to reduced learning outcomes. The ERE demonstrates that the way in which instruction is designed can be helpful for some learners but deleterious for others, so the optimal learning path for each learner may need to be different and adjust to their needs over time. Besides prior experience, other cognitive individual differences have been shown to affect how people process information during a learning episode, such as spatial ability (Hegarty et al., 2007; Johnson et al., 2022) and working memory capacity (DeCaro et al., 2008; Just & Carpenter, 1992).

What Is Adaptive Training?

Considering that learners have unique needs during the learning process, it follows that one-size-fits-all training is unlikely to meet the needs of every learner; as a result, adaptive training technology is becoming more ubiquitous across education, government, and industry sectors in an attempt to create more effective and efficient training for students and workforce. In fact, adaptive training is in such demand currently that the market for it is forecast to reach $8.8B in 2028 (Research and Markets, 2023). Adaptive training is instruction that adjusts in response to a learner's performance, skills, learning needs, ability, or other individual differences (Landsberg et al., 2012; Park & Lee, 2004; Shute & Zapata-Rivera, 2012). In other words, adaptive training is a technology-based capability that takes on the role of a human tutor and adapts instruction to address an individual learner's strengths and weaknesses. The term "adaptive training" can be considered an umbrella term that incorporates a spectrum of adaptive instruction ranging from simple to complex. On the simple end, the training could start with a pre-test, and the learner would only receive instruction about the items that they missed. On the more complex end, there are intelligent tutoring systems that modify the instruction based on a student model that the system developed from the learner's previous responses (Ma et al., 2014; Shute & Psotka, 1996). Multiple reviews of adaptive training systems have found them to be beneficial for learning by increasing learning outcomes and improving learning efficiency (Durlach & Ray, 2011; Landsberg et al., 2012; Vandewaetere et al., 2011). Despite these known benefits, adaptive training techniques have not taken off in the healthcare domain as compared to other fields, such as military and K-12 applications, which is likely in large part due to the challenges in healthcare simulation previously discussed.

ADAPTING TRAINING IN HEALTHCARE

A few key and complementary cognitive psychology concepts that underlie the success of many adaptive training systems relevant to healthcare simulation include incorporating mastery learning and deliberate practice. Mastery learning is an educational philosophy that all learners will achieve a high level of understanding (i.e., mastery) of the subject matter they are learning before moving on to new material (Bloom, 1974). Likewise, Ericsson's research in expertise revealed that what separates true experts from others is that experts engage in focused, effortful practice sessions with purpose to achieve ever higher levels of performance, which he called *deliberate practice* (Ericsson, 2004; Ericsson et al., 1993). When applied to educational settings, this would mean keeping learners within their *zone of proximal development* (ZPD; Vygotsky & Cole, 1978). Vygotsky characterized the ZPD as the difference between what learners can do on their own and what they could do with some guidance and instructional scaffolding. To achieve meaningful learning, learners need to be challenged with tasks just beyond their ability to promote optimal learning opportunities. In other words, the ZPD represents a "sweet spot" for learning that is neither too difficult nor too easy for the learner. Putting it all together, adaptive difficulty is an instructional strategy that targets these particular concepts and has been shown to be highly effective for promoting learning and performance (Wickens et al., 2013). From a healthcare simulation-based training perspective, to implement adaptive difficulty, one would provide learners exercises or scenarios that are just within their capabilities, and once they demonstrate they have mastered those exercises, they would receive more difficult ones (or start working on a new concept). This process is analogous to how medical residents train with senior clinicians, so applying these concepts within adaptive healthcare simulations would not be much of a leap conceptually.

HOW AI/ML CAN BE USED IN HEALTHCARE SIMULATION

This chapter has highlighted the challenges faced in healthcare simulation and proposed that adaptive training may be the way to address these challenges based on evidence from cognitive psychology. We discussed how healthcare simulations are typically presented as "one-size fits all" training, and that adaptive training is not well-utilized in this field. In this next section, we describe the current state of AI/ML and how specific advances in AI can help address the challenges of healthcare simulation by providing new ways to tailor training without the current limitations that simulation training faces.

What's behind the Recent AI Wave?

There is a new push to integrate AI into adaptive training, and to better understand why now is an ideal time to implement this approach in healthcare simulation, we can look at the evolution of AI that has led to exciting new possibilities. Although the fundamental mathematics and mechanics underlying artificial neural networks and how to train them have been an object of study for decades at least (Hendler, 2008), progress in deep learning was relatively slow through much of the 1980s and 1990s, decades often associated with periods that have come to be known as "AI Winters" (Hendler, 2008). However, a series of factors, each benefitting the others, began to emerge. First, advances in the availability of data, heralded by the internet and the decreasing cost of data storage, made it easier to collect large datasets that could be used to train and test algorithms. Second, hardware advances made it possible to train increasingly large models. Notably, graphics processing units (GPUs), originally developed to efficiently carry out the calculations used in computer graphics, were re-purposed to carry out the calculations used by deep learning inference and training steps. And third, AI researchers experimented with and discovered variations of neural network layers that could carry out fundamental tasks more effectively than human experts.

Each of these factors synergistically accelerated the others, leading to a renewed interest in deep learning in the late 2000s. In the sub-field of natural language processing (NLP), a similar explosion of productivity would take shape with the introduction of the *transformer* architecture (Vaswani et al., 2017). Unlike in computer vision, where input images can be standardized to all have the exact same size, NLP must deal with text that can be of arbitrary length. Larger neural networks could take larger text inputs, but the number of parameters in the network (and thus the amount of data and computational power required to train it) would often increase exponentially at best. The transformer architecture introduced a way to increase input sizes while only scaling parameter count by a quadratic factor, thus allowing for significantly larger inputs.

A sort of arms race then began, where transformer-based language models (LMs) were scaled up to larger and larger sizes, with each increase in size leading to breakthrough performances on benchmarks of language-based reasoning (Wang et al., 2019a, 2019b). Eventually, it was realized that these autoregressive LMs, operating in a generative fashion, could perform remarkably well on a variety of tasks that they were not specifically trained on (Brown et al., 2020). And thus, the era of generative LMs was launched into the public consciousness with the release of OpenAI's ChatGPT, currently based on the generative LM GPT-4 (Achiam et al., 2023).

AI for Automated Scenario Assessment

For decades now, the field of NLP has been heavily shaped by benchmark tasks and datasets, with a common complaint being that a research paper would have little chance of acceptance at a top conference unless it was able to show at least an incrementally higher performance than the best-known approach on some task. Many of these benchmark tasks were (and continue to be) based on datasets curated from human responses to some reasoning problem. For example, the influential natural language inference (NLI) task (Bowman et al., 2015) contains two sentences: a premise p and a hypothesis h, and asks a random sampling of participants to determine the inferential relationship between them (whether p implies h), in a way that captures their natural, intuitive sense of what constitutes logical consequence.

Given that so many benchmark tasks that shaped the development of LMs draw from human reasoning, it should be no surprise that LMs have some ability to not only emulate a range of human reasoning ability levels, but to distinguish between those levels as well. For this reason, an emerging body of work into the intersection of psychometrics and AI is gaining traction—both in the use of generative AI to generate psychometric test items and estimate their psychometric properties, and in the use of psychometrics to study the reasoning capabilities of AI systems. Laverghetta et al. (2021, 2022) showed that transformer-based LMs could be used to predict the item discriminability of NLI problems, but that this predictive ability differed based on the category of the problem. They later showed that generative LMs were able to create test items with surprisingly good reliability and validity, using a multi-stage prompting strategy that did not require significant fine-tuning over large datasets (Laverghetta & Licato, 2023a, 2023b). This suggests possible applications of LMs to reduce the often-costly process of employing large numbers of individuals to determine the psychometric properties of test items. Because healthcare simulation currently relies on subject matter experts to create assessments and then evaluate individuals, the utilization of LMs to distinguish the reasoning ability of learners could alleviate the challenge in healthcare training of limited human resources.

AI for Individual Differences Modeling

LMs created in recent years have shown remarkable performance on various benchmark tasks inspired by human reasoning (Wang et al., 2019b). Likewise for benchmark tasks of predicting a range of human behaviors (Brown et al., 2020). But how well do they model *individual differences* in these behaviors? As it turns out, training models in a way that optimizes aggregated behaviors of many individuals across a large benchmark dataset may produce different predictions than optimizing to predict the behaviors of one individual at a time. And the latter may be more important for adaptive simulations. For example, Beckage and Colunga (2019) used computational modeling techniques to test competing hypotheses describing language acquisition in young children. They found that although previous work (which used aggregate modeling techniques) supported one hypothesis, when the focus was on capturing language growth for individual children, a different hypothesis was supported. This suggests that the creation of effective adaptive simulation environments should use modeling techniques that take into account that predicting the behaviors of an individual may differ from techniques that are optimized toward predicting averaged behaviors of many individuals. This is particularly important to address the limitations of healthcare simulation's "one size fits all" approach, as AI-based adaptive training can be implemented to achieve optimal outcomes for the individual and not just a group of learners.

In line with the goals of adaptive training, some researchers in computational cognitive modeling have shifted focus to *individual differences modeling.* Nighojkar et al. (2022) used individual differences modeling techniques on the semantic fluency task (SFT) (Welsh et al., 1991), a simple task where participants are given a category word and asked to list as many objects that are instances of that category as they can. Using the transformer-based LM RoBERTa (Liu et al., 2019), they were able to rapidly adapt to a participant after only watching them list a few initial words and predict the words the participants would say next with a top-5 accuracy of up to 26.7%. It is interesting to note that this adaptation was done without requiring extensive training of the model or fine-tuning of the standard RoBERTa model. Although RoBERTa is a statistical LM without any *a priori* claim to be human-like, Nighojkar et al. propose a method called *hyperparameter hypothesization*, by which statistical LMs (or AI models in general) can be used to generate testable hypotheses of causal explanations of human behaviors and cognitive traits. Likewise, Fields and Licato (2023a, 2023b) applied similar individual differences modeling techniques to predict player behaviors in collectible card games. The evolution of AI models that focus on individual differences are necessary foundations for precise adaptive training models that predict and assess learner behavior during healthcare simulation.

AI for Automated Scenario Design and Adaptation

Putting the above two innovations together, it is easy to see how we can automatically design scenarios based on learning and assessment goals, rapidly adapting the parameters of the scenario to the individual using individual differences modeling, and then automatically providing high-quality assessments of the individual's performance. Going back to the Vygotskyan concept of *scaffolding*, an AI could guide the learning of a student (or simulation participant), ideally by ensuring they are exposed primarily to problems in their *ZPD* (i.e., the set of problems more difficult than those the student would be able to learn on their own, but still within their ability to learn with gentle guidance) (Shaffer & Kipp, 2014; Vygotsky, 1962, 2012; Vygotsky & Cole, 1978). However, scaffolding, and assessing whether a problem is in a student's ZPD, requires a teacher that can estimate both the difficulty of a problem, and the current competence of the student. Considering the recent advances in AI+psychometrics and individual differences modeling we have summarized earlier, implementing such a teacher into artificial simulations is now a very real possibility.

FUTURE OF AI IN HEALTHCARE SIMULATION

The diverse challenges in creating effective healthcare simulation, ranging from individualized learning content to resource constraints, underscore the need for innovation in this field. This chapter advocates for the integration of AI and ML in adaptive training to enhance healthcare education. By exploring cognitive frameworks, we highlight why specific instructional strategies are effective and should serve as the foundation for AI-based adaptive training, shaping the future of research and development in healthcare simulation.

Tailoring training to an individual typically requires significant time from a facilitator to assess the performance of a trainee and provide appropriate feedback and adaptation of learning content. Additionally, as learning outcomes are often multifaceted, there is a need for multiple dimensions of assessment, feedback, and adaptation for trainees to reach desired learning outcomes. While good instructional strategies suggest that providing prompt feedback and dynamically changing a simulation scenario based on individual performance can lead to better learning outcomes, it is difficult for a human facilitator to accurately measure and respond to trainees' performance on multiple levels during a simulation. In response to these challenges, AI solutions may address the multi-dimensional considerations needed to scale healthcare training.

In recent years, advances in AI have significantly impacted virtually every field of study, and healthcare simulation is no exception. Nevertheless, recent breakthroughs in the capabilities of state-of-the-art AI systems are poised to further change the way simulations are designed, implemented, carried out, and evaluated. In this chapter, we described the key innovations behind the recent AI wave (focusing on large LMs) and highlight some ways in which healthcare simulation technologies are, and will continue, to benefit. The future of healthcare simulation is full of potential as we investigate different ways in which AI can advance training. Although we have primarily focused on advances in natural language processing and large LMs, other modalities of generative AI are also seeing rapid progress. Image, sound, and video generation are already seeing integration into LLM products such as ChatGPT and Bard. In embracing the multifaceted potential of AI, we can meet the challenges of healthcare simulation to advance training in diverse and impactful ways.

ACKNOWLEDGMENTS

Part of this research was sponsored by the DEVCOM Analysis Center and was accomplished under Cooperative Agreement Number W911NF-22-2-0001. The views and conclusions contained in this document are those of the authors and should not be interpreted as representing the official policies, either expressed or implied, of the Army Research Office or the U.S. Government. The U.S. Government is authorized to reproduce and distribute reprints for Government purposes notwithstanding any copyright notation herein.

REFERENCES

Achiam, J., Adler, S., Agarwal, S., Ahmad, L., Akkaya, I., Aleman, F. L., …, & McGrew, B. (2023). *GPT-4 technical report*. arXiv:2303.08774.

Beckage, N. M., & Colunga, E. (2019). Network growth modeling to capture individual lexical learning. *Complexity, 2019*, 1–17.

Bloom, B. S. (1974). Time and learning. *American Psychologist, 29*(9), 682–688.

Bowman, S. R., Angeli, G., Potts, C., & Manning, C. D. (2015). A large annotated corpus for learning natural language inference. *In Conference on Empirical Methods in Natural Language Processing, EMNLP* 2015 (pp. 632–642). Association for Computational Linguistics (ACL).

Brown, T., Mann, B., Ryder, N., Subbiah, M., Kaplan, J. D., Dhariwal, P., …, & Amodei, D. (2020). Language models are few-shot learners. *Advances in Neural Information Processing Systems, 33*, 1877–1901.

Cook, D. A., Hatala, R., Brydges, R., Zendejas, B., Szostek, J. H., Wang, A. T., …, & Hamstra, S. J. (2011). Technology-enhanced simulation for health professions education: A systematic review and meta-analysis. *JAMA, 306*(9), 978–988.

DeCaro, M. S., Thomas, R. D., & Beilock, S. L. (2008). Individual differences in category learning: Sometimes less working memory capacity is better than more. *Cognition, 107*(1), 284–294.

Durlach, P. J., & Ray, J. M. (2011). *Designing Adaptive Instructional Environments: Insights from Empirical Evidence.* Technical Report 1297. Orlando, FL: US Army Research Institute for the Behavioral and Social Sciences.

Ericsson, K. A. (2004). Deliberate practice and the acquisition and maintenance of expert performance in medicine and related domains. *Academic Medicine, 79*(10), S70–S81.

Ericsson, K. A., Krampe, R. T., & Tesch-Römer, C. (1993). The role of deliberate practice in the acquisition of expert performance. *Psychological Review, 100*(3), 363–406. https://doi.org/10.1037/0033-295X.100.3.363

Fields, L., & Licato, J. (2023a, May). Player identification for collectible card games with dynamic game states. In *Proceedings of The International Florida Artificial Intelligence Research Society Conference (FLAIRS-34)* (Vol. 36).

Fields, L., & Licato, J. (2023b, October). Player identification and next-move prediction for collectible card games with imperfect information. In *Proceedings of the AAAI Conference on Artificial Intelligence and Interactive Digital Entertainment* (Vol. 19, No. 1, pp. 43–52).

Hegarty, M., Keehner, M., Cohen, C., Montello, D. R., & Lippa, Y. (2007). The Role of Spatial Cognition in Medicine: Applications for Selecting and Training Professionals. In G. Allen (Ed.), *Applied Spatial Cognition*. Mahwah, NJ: Lawrence Erlbaum Associates.

Hendler, J. (2008). Avoiding another AI winter. *IEEE Intelligent Systems, 23*(2), 2–4.

Johnson, C. I., Bailey, S. K. T., Schroeder, B. L., & Marraffino, M. D. (2022). Procedural learning in virtual reality: The role of immersion, interactivity, and spatial ability. *Technology, Mind, and Behavior, 3*(4: Winter). https://tmb.apaopen.org/pub/yfb3jhq8

Just, M. A., & Carpenter, P. A. (1992). A capacity theory of comprehension: Individual differences in working memory. *Psychological Review, 99*(1), 122–149.

Kalyuga, S. (2007). Expertise reversal effect and its implications for learner-tailored instruction. *Educational Psychology Review, 19*, 509–539.

Kalyuga, S. (2022). The Expertise Reversal Effect in Multimedia learning. In R. E. Mayer, & L. Fiorella (Eds.), *The Cambridge Handbook of Multimedia learning.* Cambridge, UK: Cambridge University Press.

Kalyuga, S., Chandler, P., Tuovinen, J., & Sweller, J. (2001). When problem solving is superior to studying worked examples. *Journal of Educational Psychology, 93*(3), 579.

Landsberg, C. R., Astwood, R. S., Van Buskirk, W. L., Townsend, L. N., Steinhauser, N. B., & Mercado, A. D. (2012). Review of adaptive training system techniques. *Military Psychology, 24*(2), 96–113.

Laverghetta, A. Jr, & Licato, J. (2023a). Automatic generation of cognitive test items using large language models. In *Proceedings of the Annual Meeting of the Psychometric Society.*

Laverghetta, A. Jr, & Licato, J. (2023b). Generating better items for cognitive assessments using large language models. In *Proceedings of the ACL 18th Workshop on Innovative Use of NLP for Building Educational Applications.*

Laverghetta, A. Jr, Nighojkar, A., Mirzakhalov, J., & Licato, J. (2021). Can transformer language models predict psychometric properties? In *Proceedings of SEM 2021: The Tenth Joint Conference on Lexical and Computational Semantics* (pp. 12–25).

Laverghetta, A. Jr, Nighojkar, A., Mirzakhalov, J., & Licato, J. (2022). Predicting human psychometric properties using computational language models. In *Proceedings of the Annual Meeting of the Psychometric Society* (pp. 151–169). Cham: Springer International Publishing.

Lioce, L., Lopreiato, J., Downing, D., Chang, T. P., Robertson, J. M., & Anderson, M., …, & Terminology and Concepts Working Group. (2020). Healthcare Simulation Dictionary (AHRQ Publication No. 20-0019). Agency for Healthcare Research and Quality.

Liu, Y., Ott, M., Goyal, N., Du, J., Joshi, M., Chen, D., Levy, O., Zettlemoyer, L., & Stoyanov, V. (2019). RoBERTa: A robustly optimized BERT pretraining approach. arXiv:1907.11692.

Ma, W., Adesope, O. O., Nesbit, J. C., & Liu, Q. (2014). Intelligent tutoring systems and learning outcomes: A meta-analysis. *Journal of Educational Psychology*, *106*(4), 901–918.

Mayer, R. E. (2020). *Multimedia learning* (3rd ed.). Cambridge, UK: Cambridge University Press.

Nighojkar, A., Khlyzova, A., & Licato, J. (2022). Cognitive modeling of semantic fluency using transformers. In *Proceedings from the IJCAI Workshop on Cognitive Aspects of Knowledge Representation*.

Okuda, Y., Bryson, E. O., DeMaria, S. Jr, Jacobson, L., Quinones, J., Shen, B., & Levine, A. I. (2009). The utility of simulation in medical education: What is the evidence? *Mount Sinai Journal of Medicine*, *76*(4), 330–343.

Park, O., & Lee, J. (2004). Adaptive Instructional Systems. In D. H. Jonassen (Ed.), *Handbook of Research for Educational Communications and Technology*, 2nd ed. (pp. 651–684). Mahwah, NJ: Lawrence Erlbaum.

Research and Markets. (2023). https://www.researchandmarkets.com/reports/5451259/global-adaptive-learning-market-2023-2028-by. Accessed 15 January 2024.

Rodziewicz, T. L., Houseman, B., & Hipskind, J. E. (2023). Medical Error Reduction and Prevention. In *StatPearls [Internet]*. Treasure Island, FL: StatPearls Publishing.

Shaffer, D. R., & Kipp, K. (2014). *Developmental Psychology: Childhood and Adolescence* (9th ed.). Belmont, CA: Wadsworth Cengage Learning.

Shute, V. J., & Psotka, J. (1996). Intelligent Tutoring Systems: Past, Present, and Future. In D. Jonassen (Ed.), *Handbook of Research for Educational Communications and Technology* (pp. 570–600). New York, NY: Macmillan.

Shute, V. J., & Zapata-Rivera, D. (2012). Adaptive educational systems. *Adaptive Technologies for Training and Education*, *7*(27), 1–35.

Vandewaetere, M., Desmet, P., & Clarebout, G. (2011). The contribution of learner characteristics in the development of computer-based adaptive learning environments. *Computers in Human Behavior*, *27*(1), 118–130.

Vaswani, A., Shazeer, N., Parmar, N., Uszkoreit, J., Jones, L., Gomez, A. N., Kaiser, L., & Polosukhin, I. (2017). Attention is all you need. Advances in neural information processing systems. In *Proceedings of the 31st International Conference on Neural Information Processing*. Long Beach, CA.

Vygotsky, L. S. (1962). *Language and thought*. Cambridge, MA: MIT Press.

Vygotsky, L. S. (2012). *Thought and language*. Cambridge, MA: MIT Press.

Vygotsky, L. S., & Cole, M. (1978). *Mind in society: Development of higher psychological processes*. Cambridge, MA: Harvard University Press.

Wang, A., Pruksachatkun, Y., Nangia, N., Singh, A., Michael, J., Hill, F., …, & Bowman, S. R. (2019a). SuperGLUE: A stickier benchmark for general-purpose language understanding systems. In *Proceedings of the 33rd International Conference on Neural Information Processing Systems* (pp. 3266–3280).

Wang, A., Singh, A., Michael, J., Hill, F., Levy, O., & Bowman, S. R. (2019b). GLUE: A multi-task benchmark and analysis platform for natural language understanding. In *7th International Conference on Learning Representations, ICLR* 2019.

Welsh, M. C., Pennington, B. F., & Groisser, D. B. (1991). A normative-developmental study of executive function: A window on prefrontal function in children. *Developmental Neuropsychology*, *7*(2), 131–149.

Wickens, C. D., Hutchins, S., Carolan, T., & Cumming, J. (2013). Effectiveness of part-task training and increasing-difficulty training strategies: A meta-analysis approach. *Human Factors*, *55*(2), 461–470.

9 Augmenting Rater Judgment Using Artificial Intelligence

Marc Cubrich, Cory Moore, Rachel T. King, and Carter Gibson

INTRODUCTION

The evaluation of human behavior and performance is ubiquitous within organizations and is critical to processes such as performance appraisal and management, assessment, and personnel selection. The most prevalent method for evaluating human behavior and performance is rating by other humans (Landy & Farr, 1980). Selection procedures, such as assessment centers, rely on the expert judgment of several trained observers. The development of this expertise is time-intensive and often requires extensive training and experience. Although raters are capable of assessing human behavior and performance with some degree of accuracy and reliability, they are prone to well-documented biases that can introduce various types of systematic and random error (Hoyt, 2000; Landy & Farr, 1980; Murphy & Balzer, 1989). Traditional rater training has focused on increasing validity and reliability while reducing rater error, but the efficacy of these interventions has also been equivocal (Bernardin & Buckley, 1981; Roch et al., 2012; Woehr & Huffcutt, 1994).

Taken together, not only is the collection of human ratings time-intensive and subject to biases, the methods for improving the accuracy of these ratings are similarly fraught with challenges. Expert rater judgment has traditionally been difficult to replicate, but advances in artificial intelligence (AI) and machine learning (ML) have made it increasingly possible to replicate human judgment with a high degree of accuracy and reliability (Campion et al., 2016; Condor, 2020). This technology has implications for augmenting the assessment of human behavior and performance through the creation of real-time decision aids to support rater judgment, and through the delivery of personalized, adaptive feedback and training. To address this opportunity, the present chapter provides a research-informed review of traditional rater training methods and highlights the ways AI can replicate expert judgment, increase efficiency, reduce bias, and improve validity and reliability. The process of developing AI models is described, along with a presentation of best practices that utilize these methods. Finally, we provide concrete recommendations for both researchers and practitioners, including model development, improving stakeholder perceptions, building trust in AI, and related ethical considerations.

DOI: 10.1201/9781032701639-9

THE ROLE AND LIMITATIONS OF SUBJECT MATTER EXPERT RATINGS

The majority of ratings made about human performance are done subjectively, that is, by other humans. This process is critical to the functioning of organizations because it has historically been one of the most effective ways of simplifying copious qualitative information into quantitative data. Once in a quantitative format, this information enables comparisons between individuals and facilitates answers to crucial questions such as whom to hire, whom to promote, and how to ensure equitable employee compensation. And despite large advances in almost every other aspect of how organizations function, human ratings are still prevalent in some of the most important decisions made in organizations, which poses several specific challenges.

At a minimum, it is important that these human ratings are consistent across raters and provide meaningful information to decision-makers. These two aspects of ratings are commonly referred to as reliability and validity, respectively. If a score given to an employment interview is not consistent across interviewers, it cannot be useful. Similarly, if it fails to provide a meaningful summary of the potential of a job candidate, even with consistency across raters, it cannot provide value to an organization. Reliability can take many forms, though for human ratings the most important is inter-rater reliability. This type of reliability refers to the extent two raters who are evaluating the same phenomenon give the same rating. This can be quantified in several ways, such as r_{WG} or Cronbach's α for quantitative data, or Cohen's κ for categorical ratings. Ultimately, the purpose of these statistics is to estimate the degree of agreement or consistency in scores (i.e., two raters may consistently be one point off from each other, but otherwise agree on ranking of ratees). Common rater training approaches (which will be described later in this chapter) often aim to increase the inter-rater reliability of raters. In organizations, one of the limitations of human raters is that it can be costly to train and maintain a pool of expert raters, not to mention issues such as rater drift that can necessitate periodic retraining.

After establishing some form of reliability, we can consider aspects of validity, which is defined as "an overall evaluative judgment for the degree to which empirical evidence and theoretical rationales support the adequacy and appropriateness of interpretation and actions based on test scores or other modes of assessment" (Messick, 1989). Basically, validity assesses whether the measurement accurately captures what you intend to measure. If a job interview is meant to provide a summary of potential if a candidate was hired, these scores can be considered valid if they predict performance. Similar to reliability, validity can be operationally defined in several ways such as criterion-related validity, where expert ratings are later compared to outcomes such as job performance or turnover, or construct validity, where scores on a particular tool are compared to existing tools that measure something similar. For a score to be useful, it needs to be *both* reliable and valid.

Given its importance to organizations and inherent limitations of subjective ratings, the topic of expert ratings has been studied, debated, and researched for well over 50 years (Guion, 1965). Over time, we've learned a significant amount regarding the specific limitations of human raters, as well as best approaches to train them. One of the largest limitations to human ratings is cognitive bias, which is a systematic and identifiable set of errors commonly made by human raters. For example, the halo effect was identified over 100 years ago (Thorndike, 1920) to describe the tendency to give high scores on unrelated aspects of performance because of something a ratee does well. That is, if you are rated on several aspects of performance such as teamwork, communication, and work quality, an effective communicator may get higher scores on other aspects. Conversely, the opposite effect may occur where several constructs receive low ratings because of one thing they do poorly, such as consistently missing deadlines leading to universally low ratings. Other biases include the similar-to-me effect, where we prefer individuals that look or think like we do, contrast effects, where somebody doing multiple raters may be biased by perceptions of previous candidates (e.g., a really high-quality candidates makes other candidates after them appear worse than they really are), or the overgeneralizing bias, where raters may make assumptions about somebody because of their group (e.g., assuming all candidates from a top business school are great candidates).

While there are a number of cognitive biases that impact human raters, the problems these biases pose are universal: they have the potential to undermine the reliability or validity of a given process. Some may be costly, such as if a White job interviewer consistently rates Black applicants lower than White ones, due to the similar-to-me effect. This type of bias could lead to lawsuits against an organization. Other biases, such as overgeneralizing, can lead to suboptimal hiring as lower quality applicants may be passed forward based solely on the reputation of the school they attended, which typically produces high-quality candidates. While steps can be taken to mitigate the impact these biases can have on decision-making, it is hard to completely overcome or eliminate them in practice. And beyond these biases, there are other limitations and downsides to the use of expert ratings.

Adding to the concerns noted above, the process of collecting human ratings of behavior and performance is often perceived as time-consuming and expensive, not only in the context of performance management but also during selection processes, including interviews and assessment centers. This perception stems from the belief that such processes are excessively subjective, resource-intensive, and ultimately unreliable, as noted above. As just one example, it is estimated that the average manager and employee spend 210 and 40 hours on performance management activities, respectively (Corporate Leadership Council, 2012). For a company of 10,000 people, this level of effort translates to a cost incurred of 30 million USD annually (Corporate Leadership Council, 2012).

Clearly, the proper assessment and evaluation of personnel require a significant investment of time and effort, which can be particularly burdensome for a single rater, let alone multiple raters, as is often required in the hiring process. This process, particularly at scale (e.g., large-scale hiring programs), can impact the efficiency and desired outcomes for organizations, such as reduced cost-per-hire or and time to offer.

Although applications of ML offer promises for gained efficiencies in this area, the development of these models has historically required considerable effort, time, and resources as well. Collecting the labeled data necessary to train, develop, and refine AI models can be similarly costly and time-intensive when relying solely on SMEs. However, an alternative approach is to leverage online crowdsourcing, which involves recruiting a large number of non-experts through online channels, rather than a small number of specialists. This method boosts efficiency by enabling the collection of vast amounts of data required for building ML models. While a single non-specialist may not individually exhibit the desired level of performance, when their responses are considered collectively, there is typically considerable convergence with ratings provided by experts (Ipeirotis et al., 2014).

Considerable advancements in Natural Language Processing (NLP) in recent years have led to notable enhancements in technologies that utilize AI and human language. These advancements have resulted in a reduced requirement for extensive training data and have consequently minimized the need for extensive human annotation to generate high-quality classification models.

While emerging technologies hold promise in enhancing and improving the process of collecting high-quality ratings of human behavior and performance, it is critical to establish a foundational understanding of traditional rater training strategies. This section will provide an overview of these strategies, highlighting their distinctions, effectiveness, and limitations. By doing so, we can better appreciate the potential impact of newer technologies in this field.

RATER TRAINING

In light of the historical context and challenges associated with SME ratings, it is imperative to explore the strategies aimed at improving the validity and reliability of human ratings. Rater training, a systematic procedure designed to improve rating accuracy by reducing rater biases (Bernardin & Buckley, 1981; Hoyt, 2000), is widely accepted as a fundamental method for improving SME ratings (Roch et al., 2012). Effective rater training is critical as it promotes fair and valid evaluations, which, in turn, play a crucial role in shaping human resource strategies and decision-making (Banks & Murphy, 1985; Landy & Farr, 1980). This section summarizes traditional rater training strategies, including rater error training (RET), performance dimension training (PDT), frame-of-reference (FOR) training, and behavioral observation training (BOT; Woehr & Huffcutt, 1994), and evaluates their effectiveness and limitations.

Traditional Rater Training Strategies

RET is a method developed to mitigate biases that influence performance evaluations, such as halo, leniency, recency effects, and attribution errors (Athey & McIntyre, 1987; Bernardin & Pence, 1980). RET emerged as a response to the prevalent psychometric errors in performance appraisal ratings, notably halo and leniency errors (Conway & Huffcutt, 1997; Woehr & Huffcutt, 1994). The training protocol educates raters about biases and equips them with strategies to circumvent these errors during performance evaluations. While evidence suggests that error training can diminish halo and leniency errors (Borman, 1979; Pulakos, 1984), it has also been critiqued for potentially compromising the overall accuracy of ratings (Bernardin & Pence, 1980; Borman, 1979). Some scholars have proposed that what are typically classified as rater errors could indeed be reflecting actual score variance (Arvey & Murphy,

1998; Hedge & Kavanagh, 1988). Moreover, there are apprehensions that error training may result in a "meaningless redistribution of ratings" (Smith, 1986). Despite these critiques, error training continues to be a widely adopted approach, although the focus on rater errors has been somewhat reduced in recent years due to a shift in the literature toward enhancing rating accuracy (Gorman et al., 2015).

PDT is a strategy that was developed as a response to the inconsistent results of RET. PDT focuses on enhancing the cognitive processing of information by raters to improve the accuracy of ratings (DeNisi et al., 1984; Feldman, 1981). In this training method, raters are educated about the specific performance dimensions that are being evaluated, including definitions and rating scales. However, feedback regarding their actual ratings is not provided. The fundamental premise of this approach is that making judgments specific to each performance dimension enhances the accuracy of the ratings (Woehr, 1992; Woehr & Huffcutt, 1994). Research generally supports the effectiveness of this training approach, but it can be influenced by various factors such as the characteristics of raters, the complexity and familiarity of the tasks being rated, and the performance dimensions being evaluated (Bernardin & Pence, 1980; Bernardin et al., 2009). Therefore, a careful design of the training program, considering these factors, is crucial for its success.

FOR training (Bernardin & Buckley, 1981) is an extension of PDT, with the addition of practice and feedback sessions. This training method involves explaining the performance dimensions, discussing behavioral examples, developing standardized evaluation rubrics, and allowing raters to make practice ratings while receiving feedback on their rating quality (DeNisi & Murphy, 2017). The goal of FOR training is to establish a shared framework that minimizes individual interpretations and biases, thereby aligning raters' understanding of performance dimensions and standards and enhancing inter-rater reliability and validity (Roch et al., 2012; Woehr, 1994). FOR training has been empirically proven to significantly enhance the accuracy of performance ratings. This is evidenced by improved inter-rater reliability and agreement when compared to control groups (Roch et al., 2012; Woehr & Huffcutt, 1994). Furthermore, numerous studies have demonstrated that FOR training can lead to improved rating accuracy (Athey & McIntyre, 1987; Bernardin & Pence, 1980; McIntyre et al., 1984; Pulakos, 1984, 1986; Schleicher & Day, 1998; Woehr, 1994).

Meta-analytic results from Woehr and Huffcutt (1994) and Roch et al. (2012) demonstrate that FOR training enhances rating accuracy, exhibiting substantial average effect sizes of $d = .83$ and $d = .50$, respectively. However, despite the proven effectiveness of FOR training, it has been critiqued for not adequately addressing the role of memory in the rating process, as it does not provide raters with strategies to process and remember behavior information for later recall, which is crucial for rating accuracy (Landy & Farr, 1980; Noonan & Sulsky, 2001; Sanchez & De La Torre, 1996). Another critique is that FOR training may lead raters to perceive certain behaviors that were not actually exhibited (Noonan & Sulsky, 2001; Sulsky & Day, 1992). Therefore, while FOR training can enhance rating accuracy, it is not a complete solution and should be used in conjunction with other strategies to improve the overall effectiveness of performance evaluations.

Lastly, BOT is a technique that was developed in response to the increasing recognition of the importance of accurate behavioral observations in ratings (Woehr & Huffcutt, 1994). This technique acknowledges the fact that raters often have to function in complex environments where they may be distracted from accurately observing performance due to multiple tasks and demands (Noonan & Sulsky, 2001). BOT typically requires raters to take notes during performance observations or to keep a record of observations over a prolonged period. The objective is to improve the accuracy of observations, thereby subsequently enhancing the accuracy of performance ratings. Studies have indicated that BOT can significantly decrease rating errors (Bernardin & Walter, 1977; Latham et al., 1975), increase the accuracy of observations (Thornton & Zorich, 1980), and improve the accuracy of ratings (Hedge & Kavanagh, 1988; Noonan & Sulsky, 2001; Pulakos, 1984). However, the effectiveness of BOT has been critiqued. Criticisms include the absence of a standard definition (Noonan & Sulsky, 2001) and practicality concerns, such as the time-intensive requirement of diaries and note-taking.

Limitations of Traditional Rater Training

Traditional rater training strategies, while effective in many respects, are not without their limitations. One of the key challenges is the sustainability of improvements and the transfer of learning from training to the actual rating process (Dierdorff et al., 2010). Brief training sessions often fail to instill lasting changes (Arthur et al., 2003), and the enhanced rating accuracy observed immediately after training frequently reverts back to baseline levels over time. Furthermore, the effects of training do not reliably generalize across various performance constructs or contexts (Arvey & Murphy, 1998). Another significant limitation is the inability of these strategies to fully address ingrained cognitive biases and limitations (Bernardin & Pence, 1980; Borman, 1979; Smith, 1986). For instance, certain approaches like FOR training do not account for issues like memory errors, perception biases, and halo effects (Bernardin et al., 2009).

The persistent human tendency toward biases like central tendency and recency poses additional challenges. Brief interventions have been unable to fundamentally change these tendencies. Lastly, the resource demands of traditional rater training limit its feasibility and scalability. Approaches like BOT and in-depth FOR training require extensive investments of time and effort. The need for ongoing reinforcement and practice poses additional burdens, hampering adoption and consistent application across organizations. While traditional rater training strategies have made positive impacts, they are constrained by inherent cognitive and practical limitations. This underscores the need for more personalized, data-driven, scalable solutions to fundamentally enhance rater competencies and improve rating validity and reliability.

Emerging AI-Enhanced Training Solutions

Feedback is a critical component of rater training programs, serving as a mechanism for raters to calibrate their judgments, correct biases, and develop their overall rating skills (Balcazar et al., 1985; London, 2003). However, traditional feedback methods have limitations in providing consistent, specific, and individualized feedback (Brett & Atwater, 2001). This presents a promising avenue for the application of AI technologies.

The emergence of cutting-edge AI technologies, such as generative AI, presents an opportunity to address these feedback limitations and enhance the rater training process. AI can potentially improve the provision of feedback by offering real-time, personalized insights for raters. By analyzing rating patterns, AI can provide comparative feedback, identify biases and errors for corrective feedback, and offer tailored suggestions for rater development. This immediate and personalized feedback could significantly enhance the effectiveness of rater training, helping raters build self-awareness and mitigate biases, thereby enhancing the overall accuracy and reliability of ratings. The emergence of AI technologies offers a promising solution to these challenges. The next section will delve deeper into the role of AI in SME ratings, exploring how AI can replicate expert judgment, reduce biases, and support rater training.

THE ROLE OF AI IN SME RATINGS

After examining traditional rater training strategies, we shift our focus to the intersection of AI and SME ratings. SMEs, with their specialized knowledge, play a pivotal role in the selection and assessment of employees, especially in analyzing candidates' natural language responses during interviews—a task increasingly being automated by AI technologies. This shift signifies a notable surge in AI adoption within organizational sciences, increasingly evident in the expanding research literature (Campion & Campion, 2023), and is reflected in the growing number of companies utilizing these technologies to enhance HR processes. In this section, we explore research highlighting AI's effectiveness in predicting SME ratings of textual responses. We also delve into how AI can be incorporated into rater training programs, potentially overcoming some of the limitations posed by traditional rater training.

Artificial Intelligence in SME Ratings: An Overview

Recent years have witnessed considerable progress in applying AI to SME ratings, a trend documented in a growing body of research (Campion & Campion, 2023; Campion et al., 2016; Hernandez & Nie, 2023; Hickman et al., 2022; Koenig et al., 2023; Landers, 2019; Putka et al., 2018; Speer et al., 2022; Thompson et al., 2023). Much of this progress has been driven by advancements in NLP, ML, and deep learning (DL), which have provided innovative methods for analyzing textual data and replicating expert evaluations and ratings. Research indicates that AI models can closely align with human SMEs in assessing candidate responses against various competencies (Koenig et al., 2023; Thompson et al., 2023). Replicating SME ratings with AI suggests potential improvements in reliability, validity, and efficiency, as well as a reduction in biases (Campion et al., 2016). This section reviews the relevant research to explore these developments.

AI techniques, especially in the domain of text analysis, offer sophisticated methods that go beyond data processing. NLP approaches such as bag-of-words (BoW) and text embeddings represent text as numbers that capture linguistic features that can be used as inputs in ML and DL models. BoW is a straightforward approach where text is represented as vectors of word occurrences (Jurafsky & Martin, 2023). Text embeddings are a more complex representation where text is converted into numerical vectors that capture deeper linguistic and semantic relationships (e.g., context and meaning; Jurafsky & Martin, 2023). These representations allow ML models to accurately predict outcomes by recognizing patterns in textual data, aligning with human evaluative judgments. A notable example of this is the study by Putka et al. (2022), which utilized BoW to estimate SME job analysis ratings. In their application, they applied BoW to process and analyze job descriptions and task statements, converting the textual content into numerical features. These features were then used to predict SME importance ratings of KSAOs for various job roles, effectively replicating the SME job analysis process. The study demonstrated a high correlation (between .74 and .84) with actual SME ratings across various KSAOs, highlighting the practicality and validity of using NLP for job analysis. The findings underscore the potential for NLP-based techniques to streamline and enhance traditional HR processes, particularly in areas like job analysis where understanding and quantifying KSAOs are crucial.

Building on the application of NLP techniques, another AI method making these advancements possible is ML. ML encompasses algorithms that enable statistical models to learn patterns in data and make predictions. Within ML, DL represents an advanced subset characterized by its neural network architectures. DL differentiates itself from traditional ML approaches by its ability to automatically learn and extract relevant features directly from raw data, thereby eliminating the need for manual feature engineering. This capability enables DL models to excel in tasks involving complex data, pattern recognition, and prediction. When coupled with text embeddings, DL models become exceptionally adept at detecting complex patterns that represent subtle linguistic nuances and semantic meanings within text. For example, ML and DL models can be trained on datasets that include SME ratings of textual responses, effectively learning and predicting how raters would likely score new responses (Campion et al., 2016; Koenig et al., 2023; Thompson et al., 2023). Consequently, these AI systems can closely replicate the scoring and evaluation patterns of SME raters who are experienced in making evaluative judgments based on language assessments (Speer et al., 2022).

Thompson et al. (2023) present another compelling case study, where DL was utilized to score the open-ended responses of pre-employment assessments. Their study analyzed job applicant data from virtual assessment centers, focusing on three distinct algorithmic methods: BoW, Long Short-Term Memory (LSTM) models, and Robustly Optimized Bidirectional Encoder Representations from Transformers (RoBERTa), a transformer-based DL model. They developed assessment items and incorporated SME ratings for job-related competencies, creating a dataset for model training. The trained models were then applied to score candidates' text responses on various competencies. Notably, the outcomes showed a high degree of alignment with human expert ratings, with RoBERTa achieving an average correlation of 0.84.

This was on par with the consensus inter-rater reliability achieved by multiple expert raters, averaging at 0.85.

Leveraging AI methods, we now have the capability to enhance a variety of assessment-related processes. This includes applications such as replicating SME ratings, improving reliability, boosting validity, increasing efficiency, mitigating biases, and automating job analysis tasks (see Table 9.1). Each of these areas showcases how AI can effectively augment and streamline the evaluation and selection process. This reflects the evolving role of AI in automating and enhancing certain tasks within assessment and selection. Although human expertise remains indispensable for many tasks, an expanding body of research suggests AI's growing potential to overtake certain scoring and evaluation tasks, often done manually by SMEs. An optimal approach seems to be a thoughtful combination of human insight and machine efficiency. AI not only brings advantages in efficiency, consistency, and bias mitigation but also serves as a valuable supplement to human judgment (Campion & Campion, 2023; Hernandez & Nie, 2023). Moving forward, the next section will explore how AI could transform rater training systems. We aim to explore a range of AI strategies, poised to enhance the effectiveness, accuracy, and fairness of rater training.

TABLE 9.1
AI Innovations in SME Ratings and Assessment Processes

Application	Description
Replicating SME Ratings	Machine learning models can be trained on datasets of assessment responses previously evaluated by SMEs. These models learn to discern patterns in the data, enabling them to predict ratings with a high degree of accuracy that often aligns closely with human SME ratings.
Improving Reliability	AI scoring models enhance reliability by applying consistent scoring rules across all assessments. Unlike human raters, AI models are not subject to fatigue, bias, or variance in judgment, leading to more consistent and reliable ratings over time.
Boosting Validity	AI models, especially those trained with accurately labeled datasets, demonstrate strong criterion-related validity. These models can capture subtle indicators within assessment responses that correlate with key performance outcomes.
Increasing Efficiency	Automated AI systems streamline the scoring and evaluation process, enabling assessment of large volumes of candidate responses. This automation significantly reduces the time and resources required, leading to cost savings and expedited decision-making in talent selection.
Mitigating Biases	AI systems can minimize biases in rating processes and predictions. By incorporating ML models that focus on multi-objective optimization, AI can consistently apply scoring rules, analyze patterns indicative of bias, and adjust to optimize for fairness alongside accuracy.
Automating Job Analysis	AI can replicate the process of SMEs in conducting job analyses by processing and analyzing job descriptions and task statements. Using NLP and ML techniques, AI can identify and quantify the importance of KSAOs, automating a traditionally labor-intensive process.

AI-Enhanced Rater Training

In light of recent advancements in AI within the domain of SME ratings, a pertinent question arises: "How might these emerging methods be effectively applied to rater training?" Recognizing the accuracy and reliability of AI models in replicating SME ratings, it's important to consider both scenarios where AI might replace human raters and where it enhances human rater training. This section introduces several innovative AI strategies aimed at augmenting rater training. These strategies, designed to improve human rater accuracy, detect and mitigate biases, and provide customized feedback, have the potential to foster development and reflective practice among rater trainees.

Rater Calibration

One potential strategy is to utilize AI models for rating calibration. This approach involves deploying a DL model, fine-tuned on a large dataset of SME ratings, to work concurrently with trainees as they practice rating textual responses. The premise is that when a significant discrepancy arises between the trainee's ratings and those predicted by the DL model, the trainee's evaluations are flagged for review. This flagging mechanism serves a dual purpose: it offers immediate feedback to trainees, drawing their attention to potential inaccuracies or biases in their assessments, and it assists in aligning the trainee's judgment with the expert judgments inherent in the training dataset. Developing such a system, however, is a complex task that entails several critical steps. First, it requires the acquisition of a high-quality, diverse dataset for fine-tuning a pre-existing, pre-trained DL model (e.g., BERT and RoBERTa). The process may also involve comparing various pre-trained models to identify the most effective one for this specific application. Following the selection and fine-tuning of the model, the next challenge is integrating it into an interactive platform. This platform should not only facilitate real-time feedback but also be user-friendly, enabling trainees to easily interpret and apply the feedback. Despite the significant engineering and data science challenges involved in creating such a system, the potential to offer real-time feedback and enhance rater calibration is apparent.

Bias Detection

Following the development of a rating calibration model and platform, another proposed strategy to enhance rater training is the development of a bias detection mechanism. This approach seeks to address a persistent limitation in traditional rater training methodologies, the detection and mitigation of biases in ratings. Central to this approach is the training of an AI model specifically designed to flag textual responses based on their susceptibility to elicit biased ratings from raters. The proposed bias susceptibility model operates by evaluating each response for specific words or phrases that are known to potentially trigger biased evaluations. The model assigns a susceptibility score to each response, reflecting the likelihood that a typical rater might respond with bias when evaluating it. This score serves as an indicator of how prone the response is to biased interpretation or judgment.

The functionality of the bias susceptibility model could be linked to the rating calibration model. When a trainee's rating significantly deviates from the prediction made by the calibration model, the bias susceptibility model provides an additional layer of feedback. It does so by analyzing the rated text and indicating if the content of the response itself might have predisposed the trainee to a biased rating. This dual-feedback mechanism—calibration deviation and bias susceptibility scoring—enables trainees to not only understand where their ratings diverge from expert patterns but also recognize and reflect on potential biases in their judgment process.

For effective implementation, the bias susceptibility model would require training and fine-tuning on a comprehensive dataset, ideally encompassing a wide array of responses that have been previously evaluated by SMEs for bias susceptibility. Through this training, the model learns to identify patterns and linguistic markers that are indicative of responses with a high potential for biased ratings. The goal is to ensure that trainees receive insightful, data-driven feedback that aids them in developing a more objective and balanced approach to their evaluations. Integrating the bias susceptibility model into the rater training process presents a method to enhance the accuracy and fairness of rater judgments. It not only aligns trainee evaluations with expert standards but also sensitizes them to the nuances of bias, fostering a more reflective and conscientious approach to rating.

Personalized Feedback

The final proposed strategy to enhance rater training through AI is the implementation of a feedback chatbot. This chatbot could be designed to provide interactive and contextually relevant feedback to trainees. The chatbot could operate by analyzing the trainee's deviations from the calibration model's predictions, the susceptibility scores from the bias detection model, the content of the textual responses being rated, and rating scales. It could synthesize this information to offer customized feedback. The primary goal is to foster a deeper understanding among trainees regarding the divergence of their ratings from the DL model's predictions, highlighting potential biases in their judgment. This tailored feedback is enhanced by the chatbot's ability for question-answering, creating an opportunity for reflective learning and development.

To implement such a chatbot, a generative AI model based on architectures like GPT could be employed to generate coherent, context-aware responses (Bommasani et al., 2022). This makes it particularly suited for engaging with trainees in a meaningful and educational manner. The chatbot could interact with trainees to clarify why their ratings deviated from model predictions, offering insights into potential biases. It could also answer related queries, thereby encouraging trainees to introspect and refine their evaluative processes. Integrating the chatbot with other strategies allows it to utilize outputs from both the rating calibration and bias detection models, ensuring a comprehensive feedback mechanism.

Implementing such an AI-driven chatbot in a real-world training environment requires meticulous design and resources, aligning with specific training needs and objectives. This system could have the potential to transform rater training, making it more interactive, insightful, and effective in honing the skills and judgment of trainees. While the development of this system is complex, it aligns with the emerging potential of AI to provide a highly personalized, efficient system for developing expertise through hands-on practice, timely feedback, and knowledge augmentation (Mollick & Mollick, 2023).

This section has explored the integration of AI in SME ratings and rater training, underscoring its emerging role in both reproducing and augmenting rater evaluations. AI is increasingly being utilized to produce ratings closely aligned with those of human SMEs, offering accuracy and efficiency in tasks such as analyzing candidates' natural language responses during interviews. These advancements in AI are making significant strides in organizational sciences and HR processes. However, while AI's role in augmenting SME ratings is substantial, it does not entirely replace the nuanced judgment of human experts. As AI technologies continue to advance, they will likely play an increasing role in training programs, offering new avenues to improve human skills. For organizations, this progression signifies a shift toward utilizing AI to improve efficiency while also preserving essential human insights. Looking ahead, the future of SME ratings and rater training lies in effectively combining AI's analytical capabilities with human expertise, aiming to achieve a more effective, accurate, and balanced assessment process.

DEVELOPING AND TRAINING AI MODELS TO REPRODUCE SME RATINGS

Developing and training AI models to reproduce SME ratings has a similar set of best practices as outlined above when building AI models to train raters. One of the most notable differences is that models built for replacing or augmenting SME ratings need to be even more accurate than those used to train raters as the cost of mistakes can be higher as discussed by the research described above.

Steps to Developing and Training a Model

The first, and arguably most important, step is collecting data to train your AI model. Any model designed to replicate human ratings can only be as effective as the data that is used to build the model. Best practices include ensuring your raters are well calibrated with one another and working from a shared mental model, ensuring that you have a diverse set of raters and/or that you are obscuring demographic information when presenting responses to be rated, and ensuring that you have response in your training data representing the full range of possible responses (e.g., if your model is being trained to provide ratings on a 1–5 scale, ensuring you have both 1's and 5's represented in your training data).

The first step to ensuring that the raters who will be rating your training data have a shared mental model is having a theoretically sound definition for the underlying construct that is being rated. For example, if raters are being asked to rate open-ended interview responses on conscientiousness, it is important to have a definition of conscientiousness to share with the raters along with a behaviorally anchored rating scale to assist raters in providing their ratings. It is also important to ensure that raters are calibrated with one another and have a shared mental model. This is where the traditional rater training strategies shared earlier in this chapter like FOR training (Bernardin & Buckley, 1981) can be used to help train raters at the start of a new project. It can also be helpful to conduct training sessions where raters all rate the same sample responses and come to a consensus on what the correct response should be (Koenig et al., 2023; Thompson et al., 2023).

Another critical decision point in building an AI model to replicate SME ratings is selecting the AI model that you intend to use. There are a number of models available, although there are a few that tend to be used more frequently for this purpose and new models are constantly being developed. Thompson et al. (2023) explored three of the most common model architectures, in order from oldest to newest, BoW, LSTM models, RoBERTa. The model that is right for a particular use case may vary, but in general newer models outperform older models as demonstrated by Thompson and colleagues. There may be future applications for the large language models that underlie generative AI here as well. A general best practice is to compare two or more appropriate models to determine which one is the best fit for your use case.

Once a model has been selected, the actual modeling work can begin. Although a deep dive on this topic is outside the scope of this chapter, at a high level, this involves actually training and fine tuning the model to predict the labeled SME ratings. There are a number of metrics that can be considered during this process. Many are more engineering and data science based, but more familiar metrics for psychologists such as correlations can be produced and examined as well. When building these models it is important to also consider other forms of reliability and validity like convergent and divergent validity and whether or not the model will produce the same score every time on the same input data (Campion et al., 2016; Speer et al., 2022).

Finally, once a model has been built and thoroughly tested, it can be deployed into the real world. This process will likely require collaborating with engineers and data scientists if they have not already been involved in the process to date. It is also important to ensure that any deployed models are being monitored on an ongoing basis to ensure that there is not model drift or any unexpected behaviors once these models are operating on real-world data rather than training data.

Considerations

There are a number of considerations that should be taken into account when building an AI model to replicate human raters. These can generally be categorized as ethical considerations, transparency and explainability, and human-AI collaboration. All of these considerations can be addressed by conducting either an internal or external AI audit to better understand how the AI tool is working (Landers & Behrend, 2022).

The most important set of considerations are ethical and legal considerations. Many ratings tasks that use human raters today have high stakes implications such as whether someone gets hired, promoted, loses a job, or passes a training exercise. Among the many ethical considerations that exist, the biggest is fairness. An AI tool that is built to replace or augment human raters will essentially scale the ratings that were used for training. If there is any intentional or unintentional bias in those ratings, it will be present in the AI tool as well. It's important to evaluate both your training data and your final model for bias or adverse impact to ensure that the algorithm being implemented is fair to everyone who is being evaluated. There are a number of ways to examine this ranging from traditional I-O methods such as the 4/5th rule to newer techniques like multi-penalty optimization (Rottman et al., 2023). Fairness is also important from a legal perspective as there are a number of laws that might apply depending on the use case of the AI rating tool and the jurisdiction in which the tool is being used.

It's also important from both an ethics and transparency perspective to ensure that the models used are as transparent and explainable as possible. Typically simpler models such as linear regression or BoW will be more explainable because they are less mathematically complicated and it is easy to see how inputs such as individual words are being weighted by the model. However, simple models do not always result in the best validity, meaning that often the models that are best equipped for the complex task of replicating human ratings are not the most explainable out of the box. There are a number of ways to address model explainability. Felzmann and colleagues (2020) propose a new transparent concept and offer a set of nine principles for designing transparent AI systems that will be discussed hereafter. Additionally, there is a subfield of ML known as explainable artificial intelligence (xAI) that includes techniques such as Local Interpretable Model-Agnostic Explanations (LIME) that can be used to open the black box of these models (Dieber & Kirrane, 2020).

The final consideration is human-AI collaboration. It is important to ensure that humans are guiding the AI development every step of the way from ensuring the raters preparing the training data are working from a shared mental model and are rating psychometrically sound concepts to the decisions that the human training the model is making when selecting a model and doing the actual training work. By having humans who are knowledgeable about the task that AI is being asked to replicate and who are knowledgeable about the necessary safeguards and compliance concerns in the loop, you can ensure that your AI rating tool is functioning as expected.

RECOMMENDATIONS

When implementing AI in the context of SME ratings, there are a number of best practices that should be kept in mind. First and foremost, the introduction of AI does not negate the need to still follow all of the best practices of our field. It is still imperative to ensure that training best practices as outlined in this chapter and elsewhere are followed and that the tool has validity, reliability, and is free of adverse impact. It can often be easy to get caught up in the technical details of building and deploying AI, but the best practices and ethical guidelines of psychology do not change just because AI has been added to one's toolbox.

It's also important to work closely with data scientists and engineers when building models (Landers, 2023). This serves a couple of purposes. By working with technical experts early in the process, it makes deploying and productionizing models easier down the line when the engineers have been involved from the beginning and have had the opportunity to give input where appropriate. It can also help ensure that the psychologists and the engineers who need to work together have a common language and that the psychologists have a seat at the table in the future when decisions are being made about technology and AI that impact our field and day-to-day work.

Stakeholder Perceptions and AI Adoption

Beyond addressing the critical practical and ethical considerations in developing AI models, it is equally important to engage in discussions concerning stakeholder perceptions, such as trust, as they can have a meaningful impact on the adoption and utilization of AI. Considering the end-user experience becomes pivotal in fostering trust and enhancing AI adoption and usage. In the context of relying on AI-produced ratings, individuals may feel uncertain about the ability of AI in producing accurate assessments compared to human judgment. Moreover, there may be a lack of understanding among individuals regarding the underlying objectives of AI in generating ratings, particularly its role in achieving fair and unbiased evaluation outcomes (Van Esch et al., 2019).

Starke et al. (2022) conducted a review of 58 studies and highlighted context-dependence of fairness perceptions. They emphasize the need for coherent theoretical frameworks and advocate for the development of reliable measures of perceived algorithmic fairness and exploration of its consequences. While applications of AI in organizations may be considered relatively nascent and ever changing, several research works began to offer theoretical frameworks and principles to understand and improve attitudes, perceptions, and behavioral outcomes of second and third parties toward automated and augmented decision-making (Felzmann et al., 2020; Langer & Landers, 2021; Mahmud et al., 2022). When seeking to foster adoption and positive reactions toward the implementation of AI in organizations, it is advisable to rely on the theoretical frameworks and principles emerging from research.

For example, "Transparency by Design" (TbD) is a set of principles offered by Felzmann et al. (2020) for automated decision-making that balances the benefits and challenges of transparency. The authors highlight the complexity of transparency, with tensions between theoretical ideals and practical implementation. They propose nine principles for designing transparent AI systems, which include considerations of relevant technical, informational, and stakeholder factors. TbD serves as a bridge between high-level AI ethics and their practical implementations, drawing inspiration from "Privacy by Design." This framework emphasizes the importance of balancing the desired level of transparency with what can actually be achieved, as excessive transparency can have adverse effects (Langer et al., 2018, 2021a). To implement TbD effectively, wider responsible design principles and stakeholder perspectives should be taken into account. Decision-makers should also seek to mitigate potential barriers to TbD implementation, such as misalignment with organizational incentives, and adopt regulatory measures to address these challenges.

Langer and Landers (2021) explore how various factors influence the attitudes, perceptions, and behavioral outcomes of second and third parties toward automated and augmented decision-making. The authors identify distinct reactions between decision augmentation, where there is human oversight, and full decision automation, where decisions are made solely by the AI system. System design choices, such as transparency, significantly affect perceptions but remain under researched. Factors affecting reactions to decision automation and augmentation are categorized into characteristics of the decision-making process, system characteristics, characteristics of second and third parties, task characteristics, and output and outcome characteristics. These factors include preferences for human control in high-stakes decisions, shifting preferences toward the system as its accuracy improves, the influence of experience, education, personality traits, and gender, the nature of the task, and the impact of system outputs on perceptions. Understanding these factors is key to building stakeholder trust in AI as a decision aid.

An ongoing evaluation of the empirical literature in this space is similarly important. Several empirical studies have contributed to our understanding of building stakeholder trust in AI as a decision aid. For example, a study by Solans et al. (2022) investigated the impact of accuracy and bias in a Decision Support System (DSS) on human performance and reliance. The results show that participants perform better when following the advice of the DSS, and the increase in score is related to both game difficulty and DSS accuracy. Participants exhibit rational behavior by adjusting their reliance on the DSS based on its accuracy. Interestingly, participants expressed moderate acceptance of the DSS in the exit survey, even when it exhibited low accuracy. These findings suggest that users may have difficulty detecting the quality of recommendations or predictions, highlighting the importance of considering user perception when deploying a DSS.

In another study by Langer et al. (2021b), the impact of Automated Decision Support Systems (ADSS) on managerial personnel selection tasks is explored. Three participant groups were studied, with one group receiving automated ranking of applicants before processing, another after processing, and a third group without any ranking. Satisfaction was higher for the support-after-processing group, and there was a notable increase in self-efficacy. However, no significant efficiency benefits were observed, possibly due to the simplicity of the tasks and participants' desire to verify the ADSS's recommendations. Psychological reactions varied based on when the support was provided, with those receiving support after processing reporting greater satisfaction and self-efficacy.

Taken together, it becomes clear that understanding and valuing stakeholder perceptions are foundational pillars for fostering adoption and trust in AI-powered tools. The journey toward successful AI implementation extends beyond accurate model development; it requires ongoing efforts to manage adoption, stakeholder reactions, and appropriate use once a tool is rolled out. Continuous engagement with end-users, along with proactive measures to address concerns and adapt to evolving needs remain critical for sustained AI integration.

Future Research

With the explosion of AI research and new generative AI models like those coming out of OpenAI, there are numerous directions for future research that are changing on an almost daily basis. In this section, we outline some possible future directions for research at the time of writing. There are likely future directions for research that will emerge as this area continues to rapidly evolve.

The biggest new frontier in AI right now is generative AI (Bommasani et al., 2022). With the creation of these new large language models, known as foundation models, AI now has capabilities beyond what anyone thought was possible even a few years ago. The most well known of these models is currently ChatGPT, the chat interface for OpenAI's foundation models (GPT3.5 and GPT4 at the time of writing). The power of these models and the ease of access for anyone with a computer have opened up new avenues for research, as well as risks if these models are being used incorrectly. For example, if a human rater tried to save time by secretly having one of these generative AI models do their ratings for them without any proper documentation or oversight this could lead to problems with the ratings if not done correctly and concerns about data security. However, these models could be used to help train raters by providing a realistic chat interface for training, and there are potentially opportunities to use these models to label data or generate BARs in a research context as well to see how well they compare to human SMEs.

Additionally, increased model explainability is an important potential area for future research. The field of xAI, as described previously in this chapter, is continuing to grow and is becoming more crucial as models become more complex and harder for humans to interpret. Ensuring that AI models that are being used to replicate human raters are as transparent and explainable as possible will go a long way toward helping to open the black box of these newer models and ensuring that they are working as intended.

Another possible direction for future research is exploring the viability of unstructured adaptive assessments. These could be assessments where instead of having to take a traditional closed ended assessment with multiple choice questions or having a high touch assessment center with multiple raters, the person being assessed could talk to a generative AI chatbot who asks questions as needed to gather data about relevant constructs and can probe for additional information until enough information has been gathered to generate a score. This type of technology is likely a long way off from being deployed, but it does present some interesting opportunities for research to explore if it is even possible and how it compares in terms of validity, reliability, time, and applicant reactions to traditional assessment methods.

CONCLUSION

In conclusion, the evaluation of human behavior and performance stands as a cornerstone within organizational processes, shaping critical functions like performance appraisal, assessment, and personnel selection. Human ratings, long relied upon for such evaluations, are not without their limitations, as they are susceptible to various biases and errors despite efforts to enhance validity and reliability through traditional rater training methods. However, the landscape is evolving with the integration of AI-powered tools and ML technologies. The integration of AI and ML presents a transformative opportunity to augment human judgment in assessing behavior and performance. By replicating expert judgment with remarkable accuracy and reliability, AI has the potential to streamline evaluation processes, reduce bias, and enhance validity and reliability. Through real-time decision aids and personalized feedback mechanisms, AI offers novel avenues for improving the efficiency and effectiveness of human rating systems.

This chapter has provided a comprehensive evaluation of traditional rater training methods and demonstrates the potential of AI to revolutionize evaluation practices. It has underscored the importance of research-informed approaches in AI model development and emphasized the need to address stakeholder perceptions and ethical considerations surrounding AI implementation. As organizations embark on the journey of integrating AI into their evaluation processes, it is imperative for researchers and practitioners alike to collaborate in developing robust AI models, navigating practical and ethical concerns, and relying upon research-driven principles and frameworks to guide this work. By embracing AI as a tool for enhancing human judgment rather than replacing it, organizations can harness its transformative potential to drive excellence in evaluation practices and ultimately, organizational success.

REFERENCES

Arthur, W., Bennett, W., Edens, P. S., & Bell, S. T. (2003). Effectiveness of training in organizations: A meta-analysis of design and evaluation features. *The Journal of Applied Psychology*, *88*(2), 234–245. https://doi.org/10.1037/0021-9010.88.2.234

Arvey, R. D., & Murphy, K. R. (1998). Performance evaluation in work settings. *Annual Review of Psychology*, *49*, 141–168. https://doi.org/10.1146/annurev.psych.49.1.141

Athey, T. R., & McIntyre, R. M. (1987). Effect of rater training on rater accuracy: Levels-of-processing theory and social facilitation theory perspectives. *Journal of Applied Psychology*, *72*(4), 567–572. https://doi.org/10.1037/0021-9010.72.4.567

Balcazar, F. E., Hopkins, B. L., & Suarez, Y. (1985). A critical, objective review of performance feedback. *Journal of Organizational Behavior Management*, *7*(3–4), 65–89. https://doi.org/10.1300/J075v07n03_05

Banks, C. G., & Murphy, K. R. (1985). Toward narrowing the research-practice gap in performance appraisal. *Personnel Psychology*, *38*(2), 335–345. https://doi.org/10.1111/j.1744-6570.1985.tb00551.x

Bernardin, H. J., & Buckley, M. R. (1981). Strategies in rater training. *The Academy of Management Review*, *6*(2), 205–212. https://doi.org/10.2307/257876

Bernardin, H. J., & Pence, E. C. (1980). Effects of rater training: Creating new response sets and decreasing accuracy. *Journal of Applied Psychology*, *65*(1), 60–66. https://doi.org/10.1037/0021-9010.65.1.60

Bernardin, H. J., Tyler, C. L., & Villanova, P. (2009). Rating level and accuracy as a function of rater personality. *International Journal of Selection and Assessment*, *17*(3), 300–310. https://doi.org/10.1111/j.1468-2389.2009.00472.x

Bernardin, H. J., & Walter, C. S. (1977). Effects of rater training and diary-keeping on psychometric error in ratings. *Journal of Applied Psychology*, *62*(1), 64–69. https://doi.org/10.1037/0021-9010.62.1.64

Bommasani, R., Hudson, D. A., Adeli, E., Altman, R., Arora, S., von Arx, S., Bernstein, M. S., Bohg, J., Bosselut, A., Brunskill, E., Brynjolfsson, E., Buch, S., Card, D., Castellon, R., Chatterji, N., Chen, A., Creel, K., Davis, J. Q., Demszky, D., …, & Liang, P. (2022). *On the Opportunities and Risks of Foundation Models* (arXiv:2108.07258). arXiv. http://arxiv.org/abs/2108.07258

Borman, W. C. (1979). Format and training effects on rating accuracy and rater errors. *Journal of Applied Psychology*, *64*(4), 410–421. https://doi.org/10.1037/0021-9010.64.4.410

Brett, J. F., & Atwater, L. E. (2001). 360 degree feedback: Accuracy, reactions, and perceptions of usefulness. *The Journal of Applied Psychology*, *86*(5), 930–942. https://doi.org/10.1037/0021-9010.86.5.930

Campion, M. A., & Campion, E. D. (2023). Machine learning applications to personnel selection: Current illustrations, lessons learned, and future research. *Personnel Psychology*, https://doi.org/10.1111/peps.12621

Campion, M. C., Campion, M. A., Campion, E. D., & Reider, M. H. (2016). Initial investigation into computer scoring of candidate essays for personnel selection. *Journal of Applied Psychology*, *101*(7), 958–975. https://doi.org/10.1037/apl0000108

Condor, A. (2020). Exploring Automatic Short Answer Grading as a Tool to Assist in Human Rating. In I. I. Bittencourt, M. Cukurova, K. Muldner, R. Luckin, & E. Millán (Eds.), *Artificial Intelligence in Education* (Vol. 12164, pp. 74–79). Springer International Publishing. https://doi.org/10.1007/978-3-030-52240-7_14

Conway, J. M., & Huffcutt, A. I. (1997). Psychometric properties of multisource performance ratings: A meta-analysis of subordinate, supervisor, peer, and self-ratings. *Human Performance*, *10*, 331–360. https://doi.org/10.1207/s15327043hup1004_2

Corporate Leadership Council. (2012). *Driving Breakthrough Performance in the New Work Environment* (Catalog No. CLC4570512SYN). Washington, DC: CEB.

DeNisi, A. S., Cafferty, T. P., & Meglino, B. M. (1984). A cognitive view of the performance appraisal process: A model and research propositions. *Organizational Behavior and Human Performance*, *33*(3), 360–396. https://doi.org/10.1016/0030-5073(84)90029-1

DeNisi, A. S., & Murphy, K. R. (2017). Performance appraisal and performance management: 100 years of progress? *Journal of Applied Psychology*, *102*(3), 421–433. https://doi.org/10.1037/apl0000085

Dieber, J., & Kirrane, S. (2020). *Why model why? Assessing the strengths and limitations of LIME* (arXiv:2012.00093). arXiv. https://doi.org/10.48550/arXiv.2012.00093

Dierdorff, E. C., Surface, E. A., & Brown, K. G. (2010). Frame-of-reference training effectiveness: Effects of goal orientation and self-efficacy on affective, cognitive, skill-based, and transfer outcomes. *Journal of Applied Psychology*, *95*(6), 1181–1191. https://doi.org/10.1037/a0020856

Feldman, J. M. (1981). Beyond attribution theory: Cognitive processes in performance appraisal. *Journal of Applied Psychology*, *66*(2), 127–148. https://doi.org/10.1037/0021-9010.66.2.127

Felzmann, H., Fosch-Villaronga, E., Lutz, C., & Tamò-Larrieux, A. (2020). Towards transparency by design for artificial intelligence. *Science and Engineering Ethics*, *26*(6), 3333–3361. https://doi.org/10.1007/s11948-020-00276-4

Gorman, C. A., Meriac, J., Ray, J., & Roddy, T. (2015). *Current Trends in Rater Training: A Survey of Rater Training Programs in American Organizations* (pp. 1–24).

Guion, R. M. (1965). *Personnel Testing*. New York: McGraw Hill.

Hedge, J. W., & Kavanagh, M. J. (1988). Improving the accuracy of performance evaluations: Comparison of three methods of performance appraiser training. *Journal of Applied Psychology*, *73*(1), 68–73. https://doi.org/10.1037/0021-9010.73.1.68

Hernandez, I., & Nie, W. (2023). The AI-IP: Minimizing the guesswork of personality scale item development through artificial intelligence. *Personnel Psychology*, *76*(4), 1101–1035. https://doi.org/10.1111/peps.12543

Hickman, L., Bosch, N., Ng, V., Saef, R., Tay, L., & Woo, S. E. (2022). Automated video interview personality assessments: Reliability, validity, and generalizability investigations. *Journal of Applied Psychology*, *107*(8), 1323–1351. https://doi.org/10.1037/apl0000695

Hoyt, W. T. (2000). Rater bias in psychological research: When is it a problem and what can we do about it? *Psychological Methods*, *5*(1), 64–86. https://doi.org/10.1037/1082-989X.5.1.64

Ipeirotis, P., Provost, F., Sheng, V., & Wang, J. (2014). Repeated labeling using multiple noisy labelers. *Data Mining and Knowledge Discovery, 28*. https://doi.org/10.1007/s10618-013-0306-1

Jurafsky, D., & Martin, J. (2023). *Speech and Language Processing* (3rd ed.). https://web.stanford.edu/~jurafsky/slp3/

Koenig, N., Tonidandel, S., Thompson, I., Albritton, B., Koohifar, F., Yankov, G., Speer, A., Hardy, III, Gibson, J. H., Frost, C., Liu, C., McNeney, M., Capman, D., Lowery, J., Kitching, S., Nimbkar, M., Boyce, A., Sun, A., Guo, T., & Newton, F. C. (2023). Improving measurement and prediction in personnel selection through the application of machine learning. *Personnel Psychology*. https://doi.org/10.1111/peps.12608

Landers, R. N. (Ed.). (2019). *The Cambridge Handbook of Technology and Employee Behavior* (1st ed.). Cambridge University Press. https://doi.org/10.1017/9781108649636

Landers, R. N. (2023). Fixing the Industrial-Organizational Psychology-Technology Interface (IOPTI). *Talent Assessment: Embracing Innovation and Mitigating Risk in the Digital Age*, 202. https://doi.org/10.1093/oso/9780197611050.003.0013

Landers, R. N., & Behrend, T. S. (2022). Auditing the AI auditors: A framework for evaluating fairness and bias in high stakes AI predictive models. *American Psychologist*. https://doi.org/10.1037/amp0000972

Landy, F. J., & Farr, J. L. (1980). Performance rating. *Psychological Bulletin, 87*, 72–107. https://doi.org/10.1037/0033-2909.87.1.72

Langer, M., Baum, K., König, C. J., Hähne, V., Oster, D., & Speith, T. (2021a). Spare me the details: How the type of information about automated interviews influences applicant reactions. *International Journal of Selection and Assessment, 29*(2), 154–169.

Langer, M., König, C. J., & Busch, V. (2021b). Changing the means of managerial work: Effects of automated decision support systems on personnel selection tasks. *Journal of Business and Psychology, 36*, 751–769.

Langer, M., König, C. J., & Fitili, A. (2018). Information as a double-edged sword: The role of computer experience and information on applicant reactions towards novel technologies for personnel selection. *Computers in Human Behavior, 81*, 19–30.

Langer, M., & Landers, R. N. (2021). The future of artificial intelligence at work: A review on effects of decision automation and augmentation on workers targeted by algorithms and third-party observers. *Computers in Human Behavior, 123*, 106878.

Latham, G. P., Wexley, K. N., & Pursell, E. D. (1975). Training managers to minimize rating errors in the observation of behavior. *Journal of Applied Psychology, 60*(5), 550–555. https://doi.org/10.1037/0021-9010.60.5.550

London, M. (2003). *Job Feedback: Giving, Seeking, and Using Feedback for Performance Improvement* (2nd ed., pp. xvi, 267). Lawrence Erlbaum Associates Publishers. https://doi.org/10.4324/9781410608871

Mahmud, H., Islam, A. N., Ahmed, S. I., & Smolander, K. (2022). What influences algorithmic decision-making? A systematic literature review on algorithm aversion. *Technological Forecasting and Social Change, 175*, 121390.

McIntyre, R. M., Smith, D. E., & Hassett, C. E. (1984). Accuracy of performance ratings as affected by rater training and perceived purpose of rating. *Journal of Applied Psychology, 69*(1), 147–156. https://doi.org/10.1037/0021-9010.69.1.147

Messick, S. (1989). Validity. In R. L. Linn (Ed.), *Educational Measurement* (3rd ed., pp. 13–103). Macmillan Publishing Co, Inc; American Council on Education.

Mollick, E. R., & Mollick, L. (2023). *Using AI to Implement Effective Teaching Strategies in Classrooms: Five Strategies, Including Prompts* (SSRN Scholarly Paper 4391243). https://doi.org/10.2139/ssrn.4391243

Murphy, K. R., & Balzer, W. K. (1989). Rater errors and rating accuracy. *Journal of Applied Psychology, 74*, 619–624. https://doi.org/10.1037/0021-9010.74.4.619

Noonan, L. E., & Sulsky, L. M. (2001). Impact of frame-of-reference and behavioral observation training on alternative training effectiveness criteria in a Canadian military sample. *Human Performance, 14*(1), 3–26. https://doi.org/10.1207/S15327043HUP1401_02

Pulakos, E. D. (1984). A comparison of rater training programs: Error training and accuracy training. *Journal of Applied Psychology, 69*(4), 581–588. https://doi.org/10.1037/0021-9010.69.4.581

Pulakos, E. D. (1986). The development of training programs to increase accuracy with different rating tasks. *Organizational Behavior and Human Decision Processes, 38*(1), 76–91. https://doi.org/10.1016/0749-5978(86)90027-0

Putka, D. J., Beatty, A. S., & Reeder, M. C. (2018). Modern prediction methods: New perspectives on a common problem. *Organizational Research Methods, 21*(3), 689–732. https://doi.org/10.1177/1094428117697041

Putka, D. J., Oswald, F. L., Landers, R. N., Beatty, A. S., McCloy, R. A., & Yu, M. C. (2022). Evaluating a natural language processing approach to estimating KSA and interest job analysis ratings. *Journal of Business and Psychology.* https://doi.org/10.1007/s10869-022-09824-0

Roch, S. G., Woehr, D. J., Mishra, V., & Kieszczynska, U. (2012). Rater training revisited: An updated meta-analytic review of frame-of-reference training. *Journal of Occupational and Organizational Psychology, 85*, 370–395. https://doi.org/10.1111/j.2044-8325.2011.02045.x

Rottman, C., Gardner, C., Liff, J., Mondragon, N., & Zuloaga, L. (2023). New strategies for addressing the diversity–validity dilemma with big data. *Journal of Applied Psychology.* https://doi.org/10.1037/apl0001084

Sanchez, J. I., & De La Torre, P. (1996). A second look at the relationship between rating and behavioral accuracy in performance appraisal. *Journal of Applied Psychology, 81*(1), 3–10. https://doi.org/10.1037/0021-9010.81.1.3

Schleicher, D. J., & Day, D. V. (1998). A cognitive evaluation of frame-of-reference rater training: Content and process issues. *Organizational Behavior and Human Decision Processes, 73*(1), 76–101. https://doi.org/10.1006/obhd.1998.2751

Smith, D. E. (1986). Training programs for performance appraisal: A review. *The Academy of Management Review, 11*(1), 22–40. https://doi.org/10.2307/258329

Solans, D., Beretta, A., Portela, M., Castillo, C., & Monreale, A. (2022). Human Response to an AI-Based Decision Support System: A User Study on the Effects of Accuracy and Bias. *arXiv preprint arXiv:2203.15514.*

Speer, A. B., Perrotta, J., Tenbrink, A. P., Wegmeyer, L. J., Delacruz, A. Y., & Bowker, J. (2022). Turning words into numbers: Assessing work attitudes using natural language processing. *Journal of Applied Psychology.* https://doi.org/10.1037/apl0001061

Starke, C., Baleis, J., Keller, B., & Marcinkowski, F. (2022). Fairness perceptions of algorithmic decision-making: A systematic review of the empirical literature. *Big Data & Society, 9*(2), 20539517221115189.

Sulsky, L. M., & Day, D. V. (1992). Frame-of-reference training and cognitive categorization: An empirical investigation of rater memory issues. *Journal of Applied Psychology, 77*(4), 501–510. https://doi.org/10.1037/0021-9010.77.4.501

Thompson, I., Koenig, N., Tonidandel, S., & Mracek, D. (2023). Deep learning in employee selection: Evaluation of algorithms to automate the scoring of open-ended assessments. *Journal of Business and Psychology, 38*(3), 509–527.

Thorndike, E. L. (1920). A constant error in psychological ratings. *Journal of Applied Psychology, 4*(1), 25–29. https://doi.org/10.1037/h0071663

Thornton, G. C., & Zorich, S. (1980). Training to improve observer accuracy. *Journal of Applied Psychology, 65*(3), 351–354. https://doi.org/10.1037/0021-9010.65.3.351

Van Esch, P., Black, J. S., & Ferolie, J. (2019). Marketing AI recruitment: The next phase in job application and selection. *Computers in Human Behavior, 90*, 215–222.

Woehr, D. J. (1992). Performance dimension accessibility: Implications for rating accuracy. *Journal of Organizational Behavior, 13*(4), 357–367. https://doi.org/10.1002/job.4030130404

Woehr, D. J. (1994). Understanding frame-of-reference training: The impact of training on the recall of performance information. *Journal of Applied Psychology, 79*(4), 525.

Woehr, D. J., & Huffcutt, A. I. (1994). Rater training for performance appraisal: A quantitative review. *Journal of Occupational and Organizational Psychology, 67*, 189–205. https://doi.org/10.1111/j.2044-8325.1994.tb00562.x

10 AI and the Employee Lifecycle

What We Know and What May Come

Ian M. Hughes and Andrew Samo

INTRODUCTION

Almost 25 years ago, *Nature* published an editorial note claiming that no original research had been published in decades because "metahumans", extremely advanced artificial intelligence (AI) agents, were conducting research so cutting-edge, so quickly that the only role left for humans was to try decoding and sharing the work of the AI (Chiang, 2000). This, of course, was not entirely true. It was a fictional piece by science fiction writer Ted Chiang, who often writes about how small technological advancements can dramatically change our lived experiences. Although this has been the only sci-fi ever published in *Nature*, much more recently *Nature*, *Science*, and thousands of other scientific journals updated their policies in very real editorial notes to prevent a new form of AI, Large Language Models (LLMs), from being published with credited authorship on research papers (Nature Editorial Board, 2023; Thorp, 2023). Today, we are at a pivotal moment where LLMs are not only tools for research but are quickly becoming integrated into the world of work, prompting us to consider the boundaries between artificial and human experiences at work (Woo et al., 2024).

The sudden emergence of LLMs as a category of generative AI models has quickly brought us closer to a reality where state-of-the-art (SOTA) models are able to productively contribute to our work. Although the deep learning architecture underlying generative AI is eight years old, it was less than two years ago that ChatGPT arrived and exploded in popularity. This popularity, perhaps due to design decisions around an accessible and intuitive chatbot-style interface, accelerated excitement and investment in generative AI models at an unprecedented rate. There were estimates that LLMs are poised to impact 80% of the workforce and 40% of total working hours (Eloundou et al., 2024) and add millions, billions, even trillions in global economic value (Goldman Sachs, 2023). In line with these grand expectations, LLMs are able to draft legal documents and interpret medical imaging with surprising accuracy, write and debug sophisticated computer code in multiple programming languages, and analyze massive amounts of unstructured data to provide actionable business insights in minutes (Guo et al., 2024).

DOI: 10.1201/9781032701639-10

Within this wave of enthusiasm and grand predictions, however, a more nuanced picture is emerging. There are mixed feelings about the role of AI at work. Alongside the hype there is apprehension, skepticism, and even aversion to the "dark side" of AI at work. These mixed feelings stem from concerns over potential bias in and ethical implications of AI decision-making, job displacement, and the alienation of the human worker from their work (Bankins & Formosa, 2023; Mikalef et al., 2022). The reality is much more tempered, however, with AI and LLMs serving as powerful augmentation tools rather than wholesale replacements for human workers. These tools are most effective when strategically deployed to enhance human capabilities and streamline the tasks that people don't want to do—LLMs can't be that bad if they are cleaning messy data sheets. As we move along Gartner's Hype Cycle, past the initial inflated expectations for ChatGPT, we're developing realistic expectations of LLM capabilities and gaining a clearer understanding of how they can be integrated into organizational processes and individual workflows. There is new empirical work, for example, indicating that these models can improve real-world call center productivity by 14–34%, boost consultant performance by 40%, and enhance skill development by 18–49% (Brynjolfsson et al., 2023, Dell'Acqua et al., 2023; Wiles et al., 2024). The purpose of this chapter is to review and consider how advancements in AI can be used to improve the experience of employees across their time at an organization—the so-called "bright side" of AI at work.

But What Exactly Is Generative AI?

It's fun and exciting to talk about AI, waxing poetic on the future of work, but what exactly is AI? Generative AI? A foundation model? A SOTA LLM? *AI* is a term broadly referring to any system designed to perform tasks typically requiring human intelligence (McCarthy et al., 1955). Human intelligence, as of recently, was thought to be uniquely human and uniquely applicable to work requiring the hallmark traits of humanity: creativity, critical thinking, and dynamic problem-solving—different from much of the repetitive work left on the assembly lines and factory floors after machine-based manufacturing and automation of the early 19th- and 20th-century industrial and scientific management revolutions.

Around the same time, the pioneers of AI were pushing the frontier of computing forward. In the 1820s, Charles Babbage built the first steam-based, mechanical calculator and in 1843 Ada Lovelace wrote the first computer program (a "For Loop") for Babbage's Analytical Engine. In doing so, Ada Lovelace also uncovered the concept of a "universal machine" which approximates today's general-purpose computers. In 1956, the Dartmouth AI Conference marked the start of our modern understanding of AI and the advent of expert systems, programmed to respond to user questions and natural language processors translating languages (sound familiar?). But in the 1980s, there was a decline in interest and investment around AI, because of increasing mistakes and rising costs due to rigid algorithms and limited memory and computational power, all leading to an AI winter. AI research was so cold during this time that scientists doing AI research reportedly came up with other names for their work. In 2017, a major technological breakthrough with Attention Mechanisms and Transformers (Vaswani et al., 2017) marked the start of an AI spring. Interest, investment, and research warmed up again and the combination of online data availability, modern computational hardware infrastructure, and Transformer architectures allowed for the scale needed for the emergence of today's generative AI systems.

Generative AI broadly refers to a type of AI that is able to learn patterns in massive amounts of raw unstructured data and then generate new outputs based on those learned statistical patterns. Generative AI can work with text, images, and audio, making it significantly more versatile and capable of handling unstructured data compared to traditional machine learning (ML) techniques requiring structured, labeled data. Part of the magic of generative AI lies in the use of embeddings and latent space. Embeddings are numerical representations capturing the associations and structures characterizing training data. By projecting data features to latent space, similar to how personality traits are abstracted to latent space, generative AI can uncover complex patterns in unstructured data and manipulate them to generate new, relevant outputs. Many generative AI use cases can benefit from both the generative outputs of these models but also from working directly with the embedding space.

Generative AI is typically built on foundation models. *Foundation models* generally refer to general-purpose AI systems that are designed to be able to effectively generate a wide variety of outputs for a range of different tasks out of the box, but that can be adapted to more specific tasks (i.e., they are foundational; Bommasani et al., 2021). Note that not all generative AI models are foundation models. Where foundation models are large and broad, there are also *narrow AI* models that are designed for a specific purpose, like translation, image recognition, or scoring personality from asynchronous video interviews (AVIs).

LLMs are popular examples of foundation models, as they are language-based general-purpose systems, deep learning models, that are trained on massive amounts of text data. By learning complex patterns in this data, these models are able to create statistically probable models of the associations between the words, semantics, and linguistic structures of text and use this contextual understanding to generate coherent, novel text (Brown et al., 2020). Although LLMs exemplify foundation models, many LLMs are also characterized by their chatbot-style interface and unique training to be helpful, honest, and harmless assistants that end up distinguishing them from basic foundation models. Reinforcement Learning from Human Feedback (RLHF; Christiano et al., 2017) is an important supervised training technique that tunes these models to respond to questions and instructions and align their outputs with human values and preferences. For example, ChatGPT is an RLHF-tuned LLM built on top of the GPT-3.5 and 4 foundation models. This structure also allows for specific ways of interacting with LLMs, such as prompting, which involves crafting natural language inputs to guide the model's reasoning and outputs. Advanced techniques like in-context learning (ICL) and chain-of-thought (CoT) prompting allow users to teach the models by example and improve their ability to understand and respond to complex tasks.

SOTA models are generative AI models that are at the frontier of AI capabilities. Although SOTA models are often extremely large, making them difficult and expensive to run locally, featuring some of the most advanced LLM capabilities, including multimodal (i.e., text and image), multilingual, math, coding, reasoning and even function calling and tool use capabilities. Some examples of SOTA models include OpenAI's GPT-4o, Anthropic's Claude 3.7 Sonnet, Meta's Llama 4 herd, Google's Gemini 2.5, DeepSeek v3 etc.—many of which are already household names.

These latter few SOTA capabilities have unlocked agentic LLM capabilities, where *LLM agents* are able to consider a problem, remember and prioritize tasks, and use tools (i.e., internet search, execute Python code, and manipulate data sheets) to achieve goals (e.g., AutoGPT and BabyAGI). SOTA and agentic models are at the cutting-edge of AI and there are new developments and breakthroughs happening that are rapidly changing the generative AI landscape. These descriptions, along with most of the content in this chapter, should be considered as a snapshot in time and subject to change.

But for now, SOTA and generative AI models are simply tools that people and organizations can use to improve their productivity and experience at work. Imagine that you had an intern on their first day of work, who can read every article written on your favorite theory, memorize every Fortune 500 employee handbook, and analyze thousands of pulse survey results—all before their morning coffee. This is the sort of capability (and first-day-at-work naivete) that generative AI models have—with 10,000 H100 GPUs, 175 billion parameters, and 10^26 FLOPs instead of coffee. Like a good intern, a LLM assistant (RLHF'd LLMs, i.e., ChatGPT, Claude, and Gemini) wants to be helpful, but may be overeager and run off in the wrong direction. You, the reader and new manager of a LLM-based intern, need to be patient, tell them what you want in a way the model understands, and guide them with examples when they are wrong. If a model is consistently off, you might be using the wrong prompt, wrong model, or wrong AI solution for your use case (try XGBoost).

Chapter Overview

In this context, the purpose of this chapter is to review and consider how advancements in AI can be used to improve the experience of employees across their time at an organization—the so-called "bright side" of AI at work. This chapter is going to focus on how AI is impacting the employee experience at work. We are going to consider how current advancements in AI being used across the employee lifecycle today, at the time of writing, and how frontier models may be used in the employee lifecycles of the future. The employee lifecycle roughly consists of the different stages an employee experiences across their time with an organization, from initial contact (i.e., organizational reputation and attraction) to departure (i.e., offboarding, exit, and advocacy), providing a structured framework for understanding how employees interact with organizations and how their psychological needs and experiences evolve over time (Beer et al., 1984). Here, for simplicity's sake, we adopt a five-stage lifecycle model, which includes (1) recruitment and selection, (2) employee onboarding and training, (3) performance management and appraisal, (4) social dynamics at work, and (5) organizational departure. The chapter is organized following this simplified framework.

Our overarching goal in writing this, as noted above, is to highlight some of the "bright side" of AI at work, review current applications and use cases, and to generate some excitement around future applications and use cases of AI at work. As organizational psychologists, this is an exciting time to be experimenting with generative AI in the workplace because these models have capabilities that allow for people analytics on steroids, personalization at scale, augmented human decision-making, adaptive interventions, and the automation of complex, creative yet routine tasks. Along with these opportunities there are a number of new research areas and also risks and ethical considerations to consider, including bias and fairness, privacy and data governance, transparency and explainability. This area moves so quickly that research and ethical guidelines often trail these technological advancements, creating a dynamic environment where scientist-practitioners must be agile. However, the human touch and perspective also remains important in this setting. As most people spend most of their waking lives at work; we as scientist-practitioners should be interested in understanding, developing, and ensuring that new technologies are human-centered and improving the experience of work for everyone.

RECRUITMENT AND SELECTION

Informed Recruitment and Selection

Job and work analysis is a foundational process informing a wide variety of HRM functions across the employee lifecycle, including recruiting, selection, performance management, and more (Sackett et al., 2023). As organizations try to attract and identify the right people, job analysis provides a systematic method for identifying the essential tasks, responsibilities, and qualifications required to effectively complete essential job tasks. The process broadly involves collecting detailed information about job tasks, work environments, and necessary knowledge, skills, abilities, and other characteristics (KSAOs) through methods such as interviews, observations, and questionnaires (Primoff, 1975), typically stemming from subject matter expert (SME) judgment. Job analysis is important as it ensures that HRM functions are relevant, comprehensive, and legally defensible. However, job analysis is also challenging, as maintaining up-to-date ratings in a constantly evolving work landscape can be burdensome and resource intensive (Bobko et al., 2008).

To help overcome some of these challenges, scientist-practitioners have started experimenting with language modeling to streamline the job analysis process. For example, Putka et al. (2023) used NLP to predict SME KSAO ratings from the language used in the associated job descriptions and task statements, finding evidence for both the validity of this approach and establishing the feasibility of language modeling in this process. Building on this work, LLMs are now being experimented with to automate the job analysis process by collecting data as a chatbot asynchronously interviewing job incumbents or helping to scrape and structure raw job details available online, identifying and mapping core tasks and competencies, and roleplaying as SMEs by creating importance ratings. Job and work analysis can be complex yet essential for subsequent HRM initiatives so a rigorous approach is necessary (Tippins et al., 2021). Breaking down the process into its modular parts (i.e., structuring data, extracting tasks and competencies, rating and mapping) and evaluating the LLMs performance on each (i.e., multistep and CoT) before streamlining the entire process is recommended.

Recruitment

Once the key characteristics of the work and the right person to do the work have been identified, organizations can start attracting and recruiting new talent. Recruitment, as an area of people science, has an image problem, where it is often viewed as an "old" and "traditional" field without much innovation (Lievens & Chapman, 2019). As recently as the early 2000s, people were still being recruited through job boards, newspaper postings, or word of mouth. As organizations strive to attract and retain top talent in an increasingly competitive landscape, LLMs appear to be revitalizing the field of recruitment by becoming capable partners in the recruiting process. This shift toward AI-enabled recruitment promises to streamline hiring procedures, reduce unconscious bias, and improve the overall candidate experience (Black & van Esch, 2020). But the increasing integration of generative AI in the recruiting process also raises important ethical considerations and challenges. As the goal of recruitment is to build and maintain high-quality talent pipelines and pools (Breaugh, 2013), ensuring that generative AI is integrated properly is quickly becoming an organizational imperative. Organizations often find themselves with an overwhelming number of job applicants for their HR team to sort through. And applicants often find themselves navigating through impersonal online application systems. However, LLMs provide an opportunity to potentially resolve both of these issues. Toward this, surveys suggest that 88% of companies in America are already using AI to enhance their recruiting processes (Laurano, 2021) and we expect this number to quickly increase in the coming years with generative AI.

There are several ways that generative AI is being used today for outreach, screening, and engagement in recruiting. First, LLMs automate talent outreach efforts, changing the ways organizations reach out to potential candidates and optimize their job postings. LLMs can create tailored job descriptions based on databases of updating role requirements, company culture descriptions, and tracked industry trends. Simultaneously, LLMs can also analyze successful job postings and recommend improvements in language, structure, and SEO to increase visibility in attracting top talent. Second, generative models are being used to streamline the applicant screening process. LLMs can quickly scan large volumes of resumes, with intelligent resume parsing, identifying key qualifications and experiences to bring qualified applicants to a human decision-maker's attention. This builds on traditional resume parsing technologies because LLMs have a greater understanding of natural language and can potentially infer potential skills and experiences not explicitly listed, due to their attention to context. For example, Hilton Hotels & Resorts implemented an AI-enabled screening system and time to hire dropped by 88% (i.e., 42 to 5 days) and L'Oreal, after implementing a similar system for resume review, reported a 90% decrease (i.e., 40 to 4 minutes) in time-to-screen-a-resume (Black & van Esch, 2020). Third, LLMs are improving talent engagement during the recruitment process, where LLM-powered chatbots are being used to maintain candidate engagement. These chatbots can provide 24/7 support, answering candidate questions immediately and potentially providing more natural, context-aware conversations throughout this initial process to potentially improve candidate experience and reduce drop-off. As the recruiting process unfolds, chatbots can also provide personalized insights to help candidates make informed decisions about job opportunities. Overall, current applications of generative AI in recruiting are meaningfully enhancing recruitment efficiency and accuracy on the organizational side and improving the candidate experience through personalization and responsiveness.

As a case study, for example, Unilever implemented an AI-powered graduate recruitment tool which has reportedly saved Unilever 100,000 hours of human recruitment time and approximately $1 million in annual recruitment costs globally. The company deployed a system that analyzes video interviews, assessing candidates' facial expressions, body language, and language use. The system is now used across Unilever's entire graduate recruitment program, with claims of increased ethnic and gender diversity in the workforce.

As organizations begin integrating LLMs into their hiring process, it's important to note that there has been a growing amount of research suggesting that people may have mixed reactions to AI in hiring, depending on several factors (see Bauer et al., 2024). This is important because the candidate experience during recruitment has been shown to significantly impact performance on pre-hiring tests, intentions to accept the job, and overall perceptions of the organization (McCarthy et al., 2017). Furthermore, the psychology underlying these acute reactions may also extend to the organization's reputation. Research has suggested that both instrumental (e.g., pay and benefits) and symbolic attributes (e.g., innovativeness and prestige) influence an organization's attractiveness as an employer (Highhouse et al., 2009). The use of AI in recruitment, specifically, could signal innovativeness and potential attract tech-savvy talent—but may also raise concerns around bias and fairness in human interaction. Mitigating negative perceptions around AI in recruitment follows many of the same recommendations as selection more broadly, including being transparent about AI use, providing explanations about what the AI is doing, and keeping the human touch in the process. In terms of bias, LLMs may help limit bias in recruitment as LLMs can be used to mask personally identifying information in resumes before they are shown to human recruiters, potentially mitigating implicit hiring biases (Tippins et al., 2021).

As we consider the future of recruiting, we may be entering an "arms race" between job applicants and recruiters, both leveraging SOTA models to gain an edge in the job market. On one side, recruiters are using AI to streamline and enhance their processes. On the other, applicants might use AI to optimize their applications, tailor their resumes, and even automatically apply for jobs that match their skills and interests. Looking ahead, we might expect several AI-driven changes to the world of recruitment: (i) automated talent pipeline building, where LLM agents play larger roles in identifying and proactively engaging potential candidates, including generating personalized outreach messages tailored to candidates' backgrounds and potential organizational fit; (ii) predictive career pathing, where LLMs could help offer candidates insights into career trajectories within the company, enhancing the appeal of long-term employment; (iii) virtual reality job previews and recruitment event may become immersive experiences where LLMs provide candidates a more realistic preview of the work environment and company culture (as well as career pathing); and blockchain-verified credentials, which sounds like a Ted Chiang story, but may become increasingly important for increasing trust in the application process as LLM agents are able to autonomously navigate hiring systems on the behalf of the truly human job applicant.

As organizations navigate the integration of AI into their recruitment processes, they must consider not only the efficiency gains but also the impact on their reputation and attractiveness as employers. By thoughtfully implementing AI technologies, companies can potentially enhance their employer brand, attract tech-savvy talent, and demonstrate their commitment to innovation and fair hiring practices. However, they must also be mindful of maintaining a human touch and addressing candidate concerns about privacy and fairness to ensure a positive overall candidate experience.

SELECTION

Once applicants have been recruited and decided to apply for the job, there is a pool of talent that the organizational decision-makers need to consider and select the right applicant from. Selection, the process of evaluating and identifying the right person for the job (Ployhart et al., 2017), is likely the part of the employee lifecycle with the most amount of research and organizational investment in AI solutions. And this trend is likely to continue with generative AI and LLMs, as scientist-practitioners are rushing to experiment with these new technologies and integrate them into people science and HRM functions (Woo et al., 2024). Fortunately, many of the approaches and principles that have been established and refined since the advent of personnel selection in WWI and WWII (Schmidt & Hunter, 1998) are applicable and transferable to the new era of generative AI. Traditional assessment focuses on assessing applicants for the KSAOs needed to succeed in the job, typically including ability, personality, and motivation (Sackett et al., 2023). In this area, although generative AI creates new opportunities for bigger data collection, streamlined measure development and refinement, and automated applicant scoring and recommendation, many of the typical KSAOs are still considered and standard validation practices for ensuring that selection systems are relevant, reliable, fair, and unbiased are still used (Landers & Behrend, 2023; Tippins et al., 2021).

Today, there are a number of exciting applications of LLMs in modern selection processes, including scoring asynchronous video interviews (AVIs), assessment centers (ACs), situational judgment tests, (SJTs), gathering biodata, and integrating interactive chatbots into the selection process:

AVIs are perhaps the current gold standard for AI-enabled selection, with surveys suggesting that over 85% of American companies are using some form of AVIs—and this number has likely only increased (Jaser et al., 2022). During AVIs, job candidates answer a series of standardized interview questions on camera and the interview responses are analyzed to score KSAOs (Hickman, Bosch, et al., 2022). For example, HireVue, an industry leading provider of AVIs, reports 90% decreases in time to hire and 50% decreases in costs for interviewing. There is a growing amount of research supporting the quality of AVIs, suggesting that they are highly effective at capturing job-relevant behaviors (Hickman et al., 2024; Koutsoumpis et al., 2024; Liff et al., 2024). And with multimodal LLMs, or LLMs with image capabilities, there are efforts toward scoring non- or para-verbal behaviors along job-relevant criteria. At the same time, there are concerns around bias—particularly with using computer vision to score nonverbal behaviors (Harris et al., 2018). Employee experience is also a challenge for AVIs because, despite increased efficiency on the hiring side, research suggests that applicants have some negative reactions to AVIs (Langer et al., 2019). Yet as AVIs become more commonplace and LLMs, text-to-speech (TTS), and speech-to-text (STT) capabilities continue to improve it seems likely that this will become a normal, seamless part of the hiring process.

ACs refer to a suite of standardized assessments (not necessarily in a physical location) designed to simulate real-world job experiences, that may include interviews, psychological tests, behavioral simulations and other exercises (e.g., leaderless discussions, in-basket or inbox activities), to observe, record, and evaluate key applicant KSAOs (Lievens et al., 2001). AC behaviors are often observed and then evaluated by multiple trained raters, which can quickly become expensive over many assessment activities and candidates. Fortunately, recent work has been exploring how AI can be used to extract and score behaviors from AC exercises. These efforts indicate that automating the scoring the verbal or text AC exercise responses with NLP and LLMs exhibits good psychometric properties and is comparable to the human scoring of ACs—with the addition of cost savings (Hickman et al., 2022).

SJTs refer to assessment methods that present job-related situations and several possible responses. Candidates are typically required to either pick the best possible response for the situation or indicate how effective or desirable to the possible responses are (Lievens et al., 2008; Motowidlo et al., 1990).[1] SJT development typically involves three key steps, including a job analysis (discussed above), identification of scenarios and responses, and creating a scoring key. At the beginning of this process, generative AI may be used to help generate new SJT situations or response option pools by using a LLM to explore the situation space identified in a job analysis and/or generating variations of response options originally identified by human experts. And narrow LLMs can be trained to facilitate scoring at the end of the process. Although there is limited public research on the validity of AI-generated SJTs, there is some evidence that LLMs are effectively able to respond to SJTs (Harwood et al., 2024).

Biodata refers to information about an applicant's job-relevant life history and experience (see Speer et al., 2022). For example, "Did you ever build a model airplane that flew" is a classic example of a biodata question for successfully selecting pilots in WWI and, similarly, asking a modern AI research scientist candidate whether they have ever built GPT-2 or written a CUDA kernel from scratch may be job relevant. LLMs have promising applications for biodata by potentially automating the identification, population of biodata inventories, and scoring from interview notes, transcriptions, or social media and trace data.

Finally, LLM-powered interactive chatbots are also being used to assess KSAOs. Chatbot assessment differs from AVIs as chatbots are dynamic, two-way interactions (Zhou et al., 2019). These interactions are fairly complex behind-the-scenes, involving the understanding of applicant natural language, active but structured communication to dynamically power the conversation, and an assessment engine to extract relevant information and score it (Jayaratne & Jayatilleke, 2020). Recent research has found evidence for the validity of chatbot-based personality assessment (see Fan et al., 2023). For example, Sapia AI offers an interactive solution that captures resume information, biodata, personality, and SJT style information within a chatbot-style interface (Jayaratne & Jayatilleke, 2020). Again, as LLMs continue to improve in their capabilities these chatbot-based assessments will also continue to improve.

Similar to traditional assessment, ensuring that the AI-powered selection tools are valid, reliable, and free from bias is essential. There are a several vectors where bias may be introduced in AI-powered selection processes. The typical process involves (i) using natural language processing, now with LLMs, to extract features from text, (ii) associating text features with observed scores from valid measures using supervised ML, and then (iii) predicting new scores from new text transcriptions. At the start of this process, a potentially overlooked issue is with bias in STT transcriptions, where transcriptions are often the raw material from AVIs, ACs, SJTs, etc., that is scored (Hickman et al., 2024). Toward the middle, there are concerns around innate bias in black-box models pretrained on WEIRD data and unreliable outputs, such as hallucinations. On the data side, there are the classic concerns around amplifying existing bias in historical data—"garbage in, garbage out" you've likely heard in your introductory stats course, which applies to posttraining too. Although hallucinations are becoming less of an issue with SOTA models, they are a necessary part of the probabilistic architecture underlying the magic of LLMs. Fortunately, many of the best practices for selection system development and validation are still applicable to LLM-based approaches (see SIOP, 2023). There are also advancements in methods for reducing subgroup differences (Campion et al., 2024; Zhang et al., 2023) and for explainability (i.e., XAI; Langer et al., 2021). For more on validation and bias in AI assessment, we refer readers to several great articles (Landers & Behrend, 2023; Langer et al., 2023; Tay et al., 2022; Tippins et al., 2021).

EMPLOYEE ONBOARDING AND TRAINING

Once selection procedures are finalized, organizations will typically enroll new hires in several onboarding and training programs. These processes are intended to socialize the new employees; that is, provide them with the knowledge, skills, attitudes, and behaviors necessary to successfully enact their formal roles within an organization (Wanberg, 2012). Onboarding and training are paramount for organizational success, with research indicating that organizations that invest in such human capital management strategies deal with less turnover and report higher worker productivity (Crook et al., 2011; Hirsch, 2017). Perhaps unsurprisingly, given the staff-hours needed to execute these programs, AI is quickly becoming an integral part of onboarding and training in many organizations.

Beginning with the former, organizations have utilized AI to both personalize and simplify the onboarding process. By training AI models on existing employee data, these technologies can provide new employees with the—often on-demand, 24/7—support (via chatbots) and institutional knowledge needed to quickly assimilate into the workforce. As a result of their accessibility, the use of such technologies often saves organizations time and money (Maheshwari, 2023). An applied example is illustrative; Automatic Data Processing (ADP), recently developed ADP Assist to assist with their human capital management practices (ADP, 2024). Relying on generative AI, machine learning techniques, and natural language processing methodologies, this software can quickly help employees find important documentation (e.g., tax forms and company policies), give them reminders about critical tasks (e.g., timesheets), and provide them with answers to job-relevant questions in seconds. From a managerial perspective, supervisors can also leverage ADP Assist to nudge employees toward desired behavior at key moments during the workday—based on that individual employee's data, as well as data from those in similar roles. Importantly, as noted by Hancock et al. (2023), AI technologies such as ADP Assist should not *replace* existing HR systems, but rather be incorporated alongside existing efforts to support HR employees.

From a training perspective, meanwhile, AI can also be used to support and supplement the interactions between trainers and trainees in the workplace. Importantly, like with onboarding processes, these systems should not replace human-to-human interaction; rather, they should be used in tandem with human efforts to acclimate employees to their roles (Ouyang & Jiao, 2021). One example of AI in the organizational training space is AI-powered learning management systems (LMSs). Akin to the recommendation algorithms in music streaming apps, AI-powered LMSs can tailor the training experience for each individual employee—adjusting the learning experience to meet a focal employee's preferences and needs (Davey, 2024). For example, if a new employee is struggling with specific material, an AI-powered LMS can draw from existing employee data in tandem with the new employee's prior activity to recommend remedial training material that can meet them where they are. Immersive, virtual reality (VR) technology powered by generative AI is also seeing widespread adoption across organizations. Indeed, industry titans like Walmart (Lewis, 2019) and the National Football League (Apstein, 2015; Harrison, 2024) have incorporated AI-powered VR into their training systems; such VR allows for trainees to hone their skills in an artificial environment high in both psychological *and* mundane realism. Further still, AI can also be used to enhance training procedures in high-risk occupations. The FBI and DEA, for example, have used "fuzzy logic" AI roleplaying systems—ones that can imitate human reasoning and cognition—to better prepare officers for civilian interactions in communities, navigating courtroom testimonies, and building rapport with suspects (Olsen, 2024). Altogether, AI can greatly benefit onboarding and training systems—as well performance management and appraisal systems, to be discussed in the upcoming section—within organizations.

Like with selection and recruitment, applying AI technologies to onboarding and training systems is not within some risks. Perhaps most notably, as hinted at throughout, overreliance on these technologies (i.e., replacing humans with them) has the potential to result in negative employee reactions (Gonzalez et al., 2022). This is especially likely if the focal employee perceives low trust with their AI-powered assistant (Choung et al., 2023; Glikson & Woolley, 2020; Schreibelmayr et al., 2023). Thus, organizations should take care to integrate AI-powered technologies slowly and carefully into their onboarding and training systems—lest they end up with a dissatisfied workforce that feels threatened by their presence (Wang et al., 2023). There are also many employees that have expressed AI-related concerns about privacy during onboarding and training procedures. Indeed, large amounts of user data is often collected during these processes—which can leave employees concerned about how those data will be used. Therefore, as AI-powered technologies are integrated into human capital management systems, organizations should be open (to the extent possible) regarding privacy-related concerns with them, and share with their employees best practices for protecting their privacy (Daniels, 2024).

There are several future directions concerning the implementation of AI-powered technologies into onboarding and training procedures in organizations. First, there appear to be efforts behind wider integration of AI mentorship programs (SHRM Advisor, 2024; Stefanic, 2024). These programs, relying on tracked user data, would allow for better matching of human mentors and mentees—in addition to providing mentees around the clock support and knowledge. These programs would, in implementation, likely look very similar to AI-powered LMSs, providing mentees with a hyper-tailored learning experience that suits their preferences, needs, and goals. From a research standpoint, more information is needed regarding employee reactions to AI-powered onboarding and training materials. Indeed, there is a (comparative) lack of quantitative, empirical evidence surrounding AI implementation at this stage of the employee lifecycle; most attention has been directed toward the intersection of AI and recruitment, selection, and assessment (Hunkenschroer & Luetge, 2022; Köchling & Wehner, 2023; Tippins et al., 2021).

PERFORMANCE MANAGEMENT AND APPRAISAL

Effective and practical performance management and appraisal systems are paramount for organizational success (Aguinis & Pierce, 2008; Murphy & Cleveland, 1995; Murphy et al., 2018). Performance appraisal refers to the—often infrequent—formal process in which employees are evaluated by a managerial figure, who assesses (and typically scores) their performance using a set of criteria before sharing their assessments with the focal employee. Performance management, meanwhile, is a term that collectively refers to the wide array of activities, policies, and procedures used by an organization to help employees improve their job performance (DeNisi & Murphy, 2017). Without effective performance appraisal and management systems, organizations will have little knowledge of who to hire, who to train, who to promote, and who to terminate. Below, we will discuss avenues for AI integration into these systems.

Performance appraisal in some organizations is already being augmented by AI-powered technologies. The primary limitation of traditional performance appraisal methods are that they are often recall-based; memories can be distorted in numerous ways (e.g., imagination inflation, see Schacter et al., 2011), which can make them unreliable in certain circumstances. By leveraging AI-powered technologies during the performance appraisal process, managerial figures can get a "full picture" look at an employee's performance between appraisal periods. Indeed, as noted by Galarza (2023), AI-powered technologies can be used to quickly process and analyze work-related data (from multiple sources, including emails and instant messages) and provide real-time feedback to employees on the job. Moreover, these technologies can also take this data, in tandem with written performance reviews from managerial figures, and create custom performance goals for individual employees (Galarza, 2023). This process of collecting employee data, providing (real-time or otherwise) feedback, and aiding with goal setting are also fundamental to performance management processes. Recently, research has started showing initial validity evidence for supervised NLP and SOTA LLMs in scoring unstructured performance appraisal text (Speer et al., 2024). Put differently, AI-powered technologies allow managers to easily engage in and implement performance management strategies that are informed by the data collected during the performance appraisal process. An applied example of an AI-augmented performance appraisal/management system is illustrative; Workday, Inc. uses generative AI—trained on a dataset that contains, on average, 625 billion employee-customer interactions per year—to create custom growth plans for individual employees, designed to embed them into their jobs and develop their skills (Workday, 2023).

Of course, there are potential limitations to consider regarding the use of AI-powered technologies during the performance appraisal and management processes. In addition to the overarching concerns surrounding privacy and negative employee reactions, there are also concerns surrounding AI-powered technologies' interpretation of employee performance data. Studies have found that certain generative AI tools, like ChatGPT, have the potential to (akin to human raters during the performance appraisal process, see Spence and Keeping, 2011; Storm et al., 2023) insert various biases into the language and content of their performance feedback (Snyder, 2023). Indeed, these technologies are often only as "good" as the data they are trained on; it is important for managers to understand that AI-powered technologies are not a panacea for performance appraisal/management woes. Put succinctly: stitching AI onto criterion-deficient performance-related systems to "fix things" is akin to putting a Band-Aid on a massive wound. There is no hard and fast work around for poorly designed performance appraisal and management systems; constructing these systems "the old-fashioned way" (i.e., via job analytic techniques) is wise before integrating AI approaches. Moving forward, researchers should continue to determine when and how AI-powered technologies produce biased performance-related output—as well as ways to eliminate such biases when they appear. Such knowledge would be of heightened importance for organizations—as would a deeper understanding of how employees evaluate and perform under AI-augmented appraisal/management systems (Brown et al., 2024).

SOCIAL DYNAMICS AT WORK AND OFFBOARDING

Apart from selection, onboarding, training, and performance management and appraisal—organizations invest a large amount of resources into ensuring that the social environment of the workplace is one that is supportive and engaging (Freier & Hughes, 2024). This is because both support and engagement are key predictors of employee performance (Mathieu et al., 2019; Mazzetti et al., 2023). Naturally, AI-powered technologies have been leveraged to promote these constructs. *Work engagement*, which is a positive, fulfilling, work-related psychological state that stems from the combination of three interrelated dimensions: vigor (i.e., energy and resilience), dedication (i.e., a sense of pride and meaning), and absorption (i.e., being happily engrossed in your work such that time flies by; Schaufeli & Bakker, 2004), is one of the more studied psychological constructs in the organizational sciences (Bakker et al., 2023; Christian et al., 2011; Knight et al., 2017). In addition to being related to increased task performance, work engagement is also related to resilience, optimism, proactivity, job satisfaction, job commitment, and even life satisfaction (Mazzetti et al., 2023). AI-powered technologies that are designed to assist with training and performance management efforts can also be leveraged to increase employee work engagement; these technologies can not only use onboarding and training data to assign people to roles that fit their interests and skills, but also break up complex tasks into smaller, more actionable parts (Hashim, 2024).

Regarding social support, one of the most widely implemented generative AI tools are chatbots. Indeed, survey data from Forbes notes that nearly half of all organizations (47%) are currently using or plan to use generative AI as digital personal assistants—providing employees with instantaneous work-related support (Haan & Watts, 2023). When employees perceive satisfactory availability of work-related support resources, they tend to report lower levels of burnout, role stress, and turnover intentions (Mathieu et al., 2019).

In the future, it is likely that AI-powered technologies will be used to not only promote engagement and provide support resources, but also to prevent burnout (Henkin, 2023). Indeed, tracking employee data (ethically and above board, as referenced in our onboarding and training section) provides AI-powered technologies the opportunity to flag employees who may be using language (in internal communications) reflective of burnout. In this case, managers would be able to provide targeted support to employees who may be experiencing such chronic stress.

Finally, the last stage of the employee lifecycle is one's departure from their organization. Also known as offboarding, AI-powered technologies can streamline procedures such as exit interviews and quickly analyze data from these and exit surveys. Akin to other stages of the employee lifecycle, these technologies can also create a compassionate and custom experience for individual employees. Specifically, generative AI systems can help employees navigate the complexities of maintaining their benefits (e.g., healthcare coverage), rolling over (or otherwise managing) their 401(k), and finding new employment (When, 2024). Moreover, benefitting those in managerial roles, these technologies can aid in finding successors for roles left vacant after employee departure. Akin to AI integration at other stages of the employee lifecycle, organizations should take great care to ensure that data collected during the offboarding process is properly protected. Moreover, considering that quality person-to-person exchanges during offboarding are often paramount for maintaining residual commitment (König et al., 2022), AI should *not* replace humans even at this final stage of the employee lifecycle. Indeed, as a throughline, the best approach is augmentation, rather than full automation.

ETHICAL CONSIDERATIONS AND CONCLUSIONS

Although there are very real concerns around bias and fairness—and even some pessimism around job alienation and displacement—with applications of generative AI and LLMs, there is also a "bright side" to a world of work integrated with these powerful new technologies. Looking forward, we can imagine a workplace where AI-powered personalized learning and development programs adapt in real-time to employees changing needs and career aspirations, organizations are using LLMs to enhance their people listening efforts and understand nuanced employee sentiment and culture to proactively boost engagement and well-being before issues arise, and generative AI becoming a partner in strategic workforce planning, providing actionable real-time insights into skill gaps, succession planning, and organizational design to ensure that every employee is aligned with organizational goals and finds their work to be purposeful.

Taking a human-centered, ethical, and responsible approach to generative AI research, development, and deployment at work is critical for ensuring that the future of AI at work is "bright". Responsible AI in the workplace must consider not only algorithmic bias, explainability, and privacy, but also fundamental ethical principles that preserve and enhance the human experience at work. Key considerations include:

1. *Human-Centric*: AI systems should augment and empower human capabilities, not replace human judgment entirely.
2. *Fairness and Non-Discrimination*: AI systems must be designed and implemented to promote equity and avoid perpetuating or exacerbating biases.
3. *Transparency, Explainability, and Accountability*: The use of AI in HR processes should be openly communicated, with clear mechanisms for explanation and redress.
4. *Privacy and Data Governance*: Robust safeguards must be in place to protect employee data and respect individual privacy rights.
5. *Continuous Monitoring and Improvement*: Regular audits and assessments should be conducted to ensure AI systems remain fair, accurate, and aligned with organizational values.

To move from principles to practice, it should also consider implementing cross-functional teams to help ensure that diverse perspectives are considered when using AI at work, establishing clear escalation pathways to question AI-driven decisions, and implementing routine responsible AI checklists and audits. There are several responsible AI principles available, including the AI Risk Management Framework from NIST in the US and the AI Act in the EU, that organizations can adopt and tailor to fit their own specific needs. To help address resistance to AI at work, organizations should prioritize transparent communication, involve employees in the design process and invest in AI upskilling, and take a phased approach to rolling out AI where it's clearly communicated how AI will augment—and not replace—existing processes. It's important to remember that human experience at work is the focus. AI should empower people to have positive work experiences, not diminish them.

For organizational psychologists, LLMs are technologically complex, but are fundamentally a new tool that simply allows us to work with large amounts of (previously) unstructured data in meaningful ways. As with any new tool, there are going to be periods of experimentation and excitement along with roadblocks and disillusionment as we realize that some of our non-LLM methods are better suited from some use cases (e.g., embedding search and RAG are exciting, but maybe SQL queries are alright for static searches). Going back to our analogy of an LLM as a highly capable-yet-naïve intern, LLMs need clear guidance and context, ethical oversight, and a human manager reviewing and validating of their outputs. So as we use generative AI across the employee lifecycle, our ultimate goal should be to create human-centered workplaces where technology enhances, rather than diminishes, the employee experience, making work more engaging, fair, and fulfilling for all.

NOTE

1. An example SJT available from SHL, an industry leading assessment provider, presents a situation where applicants are managing an understaffed, overwhelmed team but need to improve performance either by (i) setting up weekly team meetings, (ii) punishing low performers, or (iii) implementing personalized goal setting for each team member (with options one and three being preferred).

REFERENCES

ADP. (2024). *ADP Assist with Generative AI Features Makes HCM Decisions Easy, Smart and Human*. Roseland, NJ: ADP Media Center.

Aguinis, H., & Pierce, C. A. (2008). Enhancing the relevance of organizational behavior by embracing performance management research. *Journal of Organizational Behavior, 29*(1), 139–145. https://doi.org/10.1002/job.493

Apstein, S. (2015). Sports Illustrated's innovation of the year: Virtual reality. Sports Illustrated.

Bankins, S., & Formosa, P. (2023). The ethical implications of artificial intelligence (AI) for meaningful work. *Journal of Business Ethics, 185*(4), 725–740. https://doi.org/10.1007/s10551-023-05339-7

Bakker, A. B., Demerouti, E., & Sanz-Vergel, A. (2023). Job demands–resources theory: Ten years later. *Annual Review of Organizational Psychology and Organizational Behavior, 10*(1), 25–53. https://doi.org/10.1146/annurev-orgpsych-120920-053933

Bauer, T. N., Truxillo, D. M., McCarthy, J. M., & Erdogan, B. (2024). Applicant reactions to organizational recruitment processes. Essentials of Employee Recruitment, 124–144.

Beer, M. (1984). *Managing Human Assets*. New York, NY: The Free Press.

Black, J. S., & van Esch, P. (2020). AI-enabled recruiting: What is it and how should a manager use it?. *Business Horizons, 63*(2), 215–226.

Bobko, P., Roth, P. L., & Buster, M. A. (2008). A systematic approach for assessing the currency ("up-to-dateness") of job-analytic information. *Public Personnel Management, 37*(3), 261–277.

Bommasani, R., Hudson, D. A., Adeli, E., Altman, R., Arora, S., von Arx, S., Bernstein, M. S., Bohg, J., Bosselut, A., Brunskill, E., Brynjolfsson, E., Buch, S., Card, D., Castellon, R., Chatterji, N., Chen, A., Creel, K., Davis, J. Q., Demszky, D., …, Liang, P. (2021). On the opportunities and risks of foundation models. arXiv preprint arXiv:2108.07258. https://arxiv.org/abs/2108.07258

Breaugh, J. A. (2013). Employee recruitment. *Annual Review of Psychology, 64*(1), 389–416.

Brown, J., Burke, J., & Sauciuc, A. (2024). *Using Artificial Intelligence to Evaluate Employees: The Effects on Recruitment, Effort, and Retention* (SSRN Scholarly Paper 3861906). https://doi.org/10.2139/ssrn.3861906

Brown, T. B., Mann, B., Ryder, N., Subbiah, M., Kaplan, J., Dhariwal, P., Neelakantan, A., Shyam, P., Sastry, G., Askell, A., Agarwal, S., Herbert-Voss, A., Krueger, G., Henighan, T., Child, R., Ramesh, A., Ziegler, D. M., Wu, J., Winter, C., …, Sutskever, I. (2020). Language Models are Few-Shot Learners. arXiv preprint arXiv:2005.14165.

Brynjolfsson, E., Li, D., & Raymond, L. R. (2023). Generative AI at work (No. w31161). National Bureau of Economic Research. https://www.nber.org/papers/w31161

Campion, E. D., Campion, M. A., Johnson, J., Carretta, T. R., Romay, S., Dirr, B., …, & Mouton, A. (2024). Using natural language processing to increase prediction and reduce subgroup differences in personnel selection decisions. *Journal of Applied Psychology, 109*(3), 307.

Chiang, T. (2000). Catching crumbs from the table. *Nature, 405*, 517. https://doi.org/10.1038/35014679

Choung, H., David, P., & Ross, A. (2023). Trust in AI and Its Role in the Acceptance of AI Technologies. *International Journal of Human–Computer Interaction, 39*(9), 1727–1739. https://doi.org/10.1080/10447318.2022.2050543

Christian, M. S., Garza, A. S., & Slaughter, J. E. (2011). Work engagement: A quantitative review and test of its relations with task and contextual performance. *Personnel Psychology, 64*(1), 89–136. https://doi.org/10.1111/j.1744-6570.2010.01203.x

Christiano, P., Leike, J., Brown, T., Martic, M., Legg, S., & Amodei, D. (2017). Deep Reinforcement Learning from Human Preferences. In I. Guyon (Ed.), *Advances in Neural Information Processing Systems* (pp. 4299–4307). San Diego, CA: NIPS Foundation.

Crook, T. R., Todd, S. Y., Combs, J. G., Woehr, D. J., & Ketchen Jr., D. J. (2011). Does human capital matter? A meta-analysis of the relationship between human capital and firm performance. *Journal of Applied Psychology, 96*(3), 443–456. https://doi.org/10.1037/a0022147

Daniels, J. (2024). Five privacy tips for businesses in the age of AI. *Forbes.*

Davey, K. (2024). What to look for in an AI-powered LMS to improve learning. *Docebo.*

Dell'Acqua, F., McFowland III, E., Mollick, E. R., Lifshitz-Assaf, H., Kellogg, K., Rajendran, S., …, & Lakhani, K. R. (2023). Navigating the jagged technological frontier: Field experimental evidence of the effects of AI on knowledge worker productivity and quality. Harvard Business School Technology & Operations Mgt. Unit Working Paper, (24-013). SSRN: https://ssrn.com/abstract=4573321 or https://doi.org/10.2139/ssrn.4573321

DeNisi, A. S., & Murphy, K. R. (2017). Performance appraisal and performance management: 100 years of progress? *Journal of Applied Psychology, 102*(3), 421–433. https://doi.org/10.1037/apl0000085

Eloundou, T., Manning, S., Mishkin, P., & Rock, D. (2024). GPTs are GPTs: Labor market impact potential of LLMs. *Science, 384*(702), 1306–1307. https://doi.org/10.1126/science.adj0998

Fan, J., Sun, T., Liu, J., Zhao, T., Zhang, B., Chen, Z., …, & Hack, E. (2023). How well can an AI chatbot infer personality? Examining psychometric properties of machine-inferred personality scores. *Journal of Applied Psychology, 108*(8), 1277.

Freier, L. M., & Hughes, I. M. (2024). Promoting Well-Being and Innovation in Startups: The Role of the Social Environment. In N. Blacksmith & M. E. McCusker (Eds.), *Data-Driven Decision Making in Entrepreneurship.* Boca Raton, FL: CRC Press.

Galarza, A. (2023). Revolutionizing performance reviews with generative AI. Forbes.

Glikson, E., & Woolley, A. W. (2020). Human trust in artificial intelligence: Review of empirical research. *Academy of Management Annals, 14*(2), 627–660. https://doi.org/10.5465/annals.2018.0057

Goldman Sachs. (2023). Generative AI could raise global GDP by 7%. https://www.goldmansachs.com/insights/articles/generative-ai-could-raise-global-gdp-by-7-percent.html

Gonzalez, M., Liu, W., Shirase, L., Tomczak, D., Lobbe, C., Justenhoven, R., & Martin, N. (2022). Allying with AI? Reactions toward human-based, AI/ML-based, and augmented hiring processes. *Computers in Human Behavior, 130*, 107179. https://doi.org/10.1016/j.chb.2022.107179

Guo, F., Gallagher, C. M., Sun, T., Tavoosi, S., & Min, H. (2024). Smarter people analytics with organizational text data: Demonstrations using classic and advanced NLP models. *Human Resource Management Journal, 34*(1), 39–54.

Haan, K., & Watts, R. (2023). *How businesses are using artificial intelligence in 2024.*

Hancock, B., Schaninger, B., & Yee, L. (2023). How generative AI could support—Not replace—Human resources. McKinsey.

Harris, K. D., Murray, P., & Warren, E. (2018). Letter to U.S. Equal Employment Opportunity Commission regarding risks of facial recognition technology. Retrieved from https://www.scribd.com/document/388920670/SenHarris-EEOC-Facial-Recognition-2

Harrison, D. (2024). Commanders QB Jayden Daniels describes how virtual reality helps him. Sports Illustrated.

Harwood, H., Roulin, N., & Iqbal, M. Z. (2024). "Anything you can do, I can do": Examining the use of ChatGPT in situational judgement tests for professional program admission. *Journal of Vocational Behavior, 154*, 104013.

Hashim, S. (2024). Artificial intelligence at work: Enhancing employee engagement and business success. *Harvard Business Review.* https://hbr.org/sponsored/2024/01/artificial-intelligence-at-work-enhancing-employee-engagement-and-business-success

Henkin, D. (2023). Combating employee burnout with AI and future of work policies. *Forbes.*

Hickman, L., Bosch, N., Ng, V., Saef, R., Tay, L., & Woo, S. E. (2022). Automated video interview personality assessments: Reliability, validity, and generalizability investigations. *Journal of Applied Psychology, 107*(8), 1323–1351.

Hickman, L., Langer, M., Saef, R. M., & Tay, L. (2024). Automated speech recognition bias in personnel selection: The case of automatically scored job interviews.

Highhouse, S., Brooks, M. E., & Gregarus, G. (2009). An organizational impression management perspective on the formation of corporate reputations. *Journal of Management, 35*(6), 1481–1493.

Hirsch, A. S. (2017). Don't underestimate the importance of good onboarding. *SHRM.*

Hunkenschroer, A. L., & Luetge, C. (2022). Ethics of AI-enabled recruiting and selection: A review and research agenda. *Journal of Business Ethics, 178*(4), 977–1007. https://doi.org/10.1007/s10551-022-05049-6

Jaser, Z., Petrakaki, D., Starr, R., & Oyarbide-Magaña, E. (2022, January 27). Where automated job interviews fall short. *Harvard Business Review.* https://hbr.org/2022/01/where-automated-job-interviews-fall-short

Jayaratne, M., & Jayatilleke, B. (2020). Predicting personality using answers to open-ended interview questions. *IEEE Access, 8*, 115345–115355.

Knight, C., Patterson, M., & Dawson, J. (2017). Building work engagement: A systematic review and meta-analysis investigating the effectiveness of work engagement interventions. *Journal of Organizational Behavior, 38*(6), 792–812. https://doi.org/10.1002/job.2167

Köchling, A., & Wehner, M. C. (2023). Better explaining the benefits why AI? Analyzing the impact of explaining the benefits of AI-supported selection on applicant responses. *International Journal of Selection and Assessment, 31*(1), 45–62. https://doi.org/10.1111/ijsa.12412

König, C. J., Richter, M., & Isak, I. (2022). Exit interviews as a tool to reduce parting employees' complaints about their former employer and to ensure residual commitment. *Management Research Review, 45*(3), 381–397. https://doi.org/10.1108/MRR-02-2021-0148

Koutsoumpis, A., Ghassemi, S., Oostrom, J. K., Holtrop, D., Van Breda, W., Zhang, T., & de Vries, R. E. (2024). Beyond traditional interviews: Psychometric analysis of asynchronous video interviews for personality and interview performance evaluation using machine learning. *Computers in Human Behavior, 154*, 108128. https://doi.org/10.1016/j.chb.2023.108128

Landers, R. N., & Behrend, T. S. (2023). Auditing the AI auditors: A framework for evaluating fairness and bias in high stakes AI predictive models. *American Psychologist, 78*(1), 36.

Langer, M., Oster, D., Speith, T., Hermanns, H., Kästner, L., Schmidt, E., …, & Baum, K. (2021). What do we want from Explainable Artificial Intelligence (XAI)?–A stakeholder perspective on XAI and a conceptual model guiding interdisciplinary XAI research. *Artificial Intelligence, 296*, 103473.

Langer, M., Roulin, N., & Oostrom, J. K. (2023). Diversity and technology—Challenges for the next decade in personnel selection. *International Journal of Selection & Assessment, 31*(3), 355–360.

Langer, M., König, C. J., & Papathanasiou, M. (2019). Highly automated job interviews: Acceptance under the influence of stakes. *International Journal of Selection and Assessment, 27*(3), 217–234.

Laurano, M. (2021). The power of AI in talent acquisition. Aptitude Research Report. https://www.aptituderesearch.com/wp-content/uploads/2022/03/Apt_PowerofAI_Report-0322_Rev4.pdf

Lievens, F., & Chapman, D. (2019). Recruitment and selection. The SAGE handbook of human resource management, 123–150.

Lievens, F., Klimoski, R. J., Cooper, C. L., & Robertson, I. T. (2001). Understanding the assessment center process: Where are we now.

Lievens, F., Peeters, H., & Schollaert, E. (2008). Situational judgment tests: A review of recent research. *Personnel Review, 37*(4), 426–441.

Liff, J., Mondragon, N., Gardner, C., Hartwell, C. J., & Bradshaw, A. (2024). Psychometric properties of automated video interview competency assessments. *Journal of Applied Psychology, 109*(6), 921–948. https://doi.org/10.1037/apl0001173

Lewis, N. (2019). Walmart revolutionizes its training with virtual reality. SHRM.

Maheshwari, R. (2023). Advantages of artificial intelligence (AI) in 2024. Forbes.

Mathieu, M., Eschleman, K. J., & Cheng, D. (2019). Meta-analytic and multiwave comparison of emotional support and instrumental support in the workplace. *Journal of Occupational Health Psychology, 24*(3), 387–409. https://doi.org/10.1037/ocp0000135

Mazzetti, G., Robledo, E., Vignoli, M., Topa, G., Guglielmi, D., & Schaufeli, W. B. (2023). Work engagement: A meta-analysis using the job demands-resources model. *Psychological Reports, 126*(3), 1069–1107. https://doi.org/10.1177/00332941211051988

McCarthy, J., Minsky, M. L., Rochester, N., & Shannon, C.E. (1955). A proposal for the Dartmouth Summer Research Project on Artificial Intelligence.

McCarthy, J. M., Bauer, T. N., Truxillo, D. M., Anderson, N. R., Costa, A. C., & Ahmed, S. M. (2017). Applicant perspectives during selection: A review addressing "So what?," "What's new?," and "Where to next?". *Journal of Management*, *43*(6), 1693–1725.

Mikalef, P., Conboy, K., Lundström, J. E., & Popovič, A. (2022). Thinking responsibly about responsible AI and "the dark side" of AI. *European Journal of Information Systems*, *31*(3), 257–268. https://doi.org/10.1080/0960085X.2022.2026621

Motowidlo, S. J., Dunnette, M. D., & Carter, G. W. (1990). An alternative selection procedure: The low-fidelity simulation. *Journal of Applied Psychology*, *75*(6), 640–647.

Murphy, K. R., & Cleveland, J. N. (1995). *Understanding Performance Appraisal: Social, Organizational, and Goal-Based Perspectives.* Thousand Oaks, CA: SAGE Publications, Inc.

Murphy, K. R., Cleveland, J. N., & Hanscom, M. E. (2018). *Performance Appraisal and Management.* Thousand Oaks, CA: SAGE Publications Inc.

Nature Editorial Board. (2023, January 24). Tools such as ChatGPT threaten transparent science; here are our ground rules for their use. *Nature.* https://doi.org/10.1038/d41586-023-00191-1

Olsen, D. (2024). Fuzzy logic AI used to train police and help make Americans safer. SIMmersion.

Ouyang, F., & Jiao, P. (2021). Artificial intelligence in education: The three paradigms. *Computers and Education: Artificial Intelligence*, *2*, 100020. https://doi.org/10.1016/j.caeai.2021.100020

Ployhart, R. E., Schmitt, N., & Tippins, N. T. (2017). Solving the supreme problem: 100 years of selection and recruitment at the Journal of Applied Psychology. *Journal of Applied Psychology*, *102*(3), 291.

Primoff, E. S. (1975). How to prepare and conduct job element examinations (Vol. *75*, No. 1). US Civil Service Commission, Personnel Research and Development Center.

Putka, D. J., Oswald, F. L., Landers, R. N., Beatty, A. S., McCloy, R. A., & Yu, M. C. (2023). Evaluating a natural language processing approach to estimating KSA and interest job analysis ratings. *Journal of Business and Psychology*, *38*(2), 385–410.

Sackett, P. R., Zhang, C., Berry, C. M., & Lievens, F. (2023). Revisiting the design of selection systems in light of new findings regarding the validity of widely used predictors. *Industrial and Organizational Psychology*, *16*(3), 283–300.

Schacter, D. L., Guerin, S. A., & Jacques, P. L. S. (2011). Memory distortion: An adaptive perspective. *Trends in Cognitive Sciences*, *15*(10), 467–474. https://doi.org/10.1016/j.tics.2011.08.004

Schaufeli, W. B., & Bakker, A. B. (2004). Job demands, job resources, and their relationship with burnout and engagement: A multi-sample study. *Journal of Organizational Behavior*, *25*(3), 293–315. https://doi.org/10.1002/job.248

Schreibelmayr, S., Moradbakhti, L., & Mara, M. (2023). First impressions of a financial AI assistant: Differences between high trust and low trust users. *Frontiers in Artificial Intelligence*, *6*. https://doi.org/10.3389/frai.2023.1241290

Schmidt, F. L., & Hunter, J. E. (1998). The validity and utility of selection methods in personnel psychology: Practical and theoretical implications of 85 years of research findings. *Psychological Bulletin*, *124*(2), 262.

SHRM Advisor. (2024). Gen AI mentorship: Guiding employees on the path to excellence. *SHRM.*

SIOP. (2023). Considerations and recommendations for the validation and use of AI-based assessments for employee selection. https://www.siop.org/Portals/84/SIOP%20Considerations%20and%20Recommendations%20for%20the%20Validation%20and%20Use%20of%20AI-Based%20Assessments%20for%20Employee%20Selection%20010323.pdf?ver=5w576kFXzxLZNDMoJqdIMw%3d%3d

Snyder, K. (2023). ChatGPT writes performance feedback. *Textio.*

Speer, A. B., Tenbrink, A. P., Wegmeyer, L. J., Sendra, C. C., Shihadeh, M., & Kaur, S. (2022). Meta-analysis of biodata in employment settings: Providing clarity to criterion and construct-related validity estimates. *Journal of Applied Psychology, 107*(10), 1678.

Speer, A. B., Perrotta, J., & Kordsmeyer, T. L. (2024). Taking it easy: Off-the-shelf versus fine-tuned supervised modeling of performance appraisal text. *Organizational Research Methods*, 10944281241271249.

Spence, J. R., & Keeping, L. (2011). Conscious rating distortion in performance appraisal: A review, commentary, and proposed framework for research. *Human Resource Management Review, 21*(2), 85–95. https://doi.org/10.1016/j.hrmr.2010.09.013

Stefanic, D. (2024, June 19). AI-Powered Mentorship Programs. *Hyperspace^mv - the Metaverse for Business Platform*. https://hyperspace.mv/ai-mentorship/

Storm, K. I. L., Reiss, L. K., Guenther, E. A., Clar-Novak, M., & Muhr, S. L. (2023). Unconscious bias in the HRM literature: Towards a critical-reflexive approach. *Human Resource Management Review, 33*(3), 100969. https://doi.org/10.1016/j.hrmr.2023.100969

Tay, L., Woo, S. E., Hickman, L., Booth, B. M., & D'Mello, S. (2022). A conceptual framework for investigating and mitigating machine-learning measurement bias (MLMB) in psychological assessment. *Advances in Methods and Practices in Psychological Science, 5*(1), Article 25152459211061337. https://doi.org/10.1177/25152459211061337

Thorp, H. H. (2023). ChatGPT is fun, but not an author. *Science, 379*(6630), 313. https://doi.org/10.1126/science.adg7879

Tippins, N., Oswald, F., & McPhail, S. (2021). Scientific, legal, and ethical concerns about AI-based personnel selection tools: A call to action. *Personnel Assessment and Decisions, 7*(2). https://doi.org/10.25035/pad.2021.02.001

Vaswani, A., Shazeer, N., Parmar, N., Uszkoreit, J., Jones, L., Gomez, A. N., Kaiser, Ł., & Polosukhin, I. (2017). Attention is all you need. In I. Guyon, U. V. Lux-burg, S. Bengio, H. Wallach, R. Fergus, S. Vishwanathan, R. Gar-nett (Eds.), *Advances in Neural Information Processing Systems* (pp. 5998–6008). Red Hook, NY: Curran Associates.

Wanberg, C. (2012). *The Oxford Handbook of Organizational Socialization*. New York, NY: Oxford University Press.

Wang, X., Li, L., Tan, S. C., Yang, L., & Lei, J. (2023). Preparing for AI-enhanced education: Conceptualizing and empirically examining teachers' AI readiness. *Computers in Human Behavior, 146*, 107798. https://doi.org/10.1016/j.chb.2023.107798

When. (2024). When announces $4.6 million in seed funding to transform the employee offboarding experience. PR Newswire.

Wiles, E., Krayer, L., Abbadi, M., Awasthi, U., Kennedy, R., Mishkin, P., …, & Candelon, F. (2024). GenAI as an Exoskeleton: Experimental Evidence on Knowledge Workers Using GenAI on New Skills. SSRN: https://papers.ssrn.com/sol3/papers.cfm?abstract_id=4944588

Woo, S. E., Tay, L., & Oswald, F. (2024). Artificial intelligence, machine learning, and big data: Improvements to the science of people at work and applications to practice. *Personnel Psychology, 77*(4), 1387–1402.

Workday. (2023). Workday unveils new generative AI capabilities to amplify human performance at work. PR Newswire.

Zhang, N., Wang, M., Xu, H., Koenig, N., Hickman, L., Kuruzovich, J., …, & Kim, Y. (2023). Reducing subgroup differences in personnel selection through the application of machine learning. *Personnel Psychology, 76*(4), 1125–1159.

Zhou, M. X., Chen, W., Xiao, Z., Yang, H., Chi, T., & Williams, R. (2019, March). Getting virtually personal: chatbots who actively listen to you and infer your personality. In *Companion Proceedings of the 24th International Conference on Intelligent User Interfaces* (pp. 123–124).

Index

Note: Page numbers in **bold** denote tables and page numbers in *italics* denote figures.

A

B

C

D

G

H

I

J

K

L

M

N

O

P

R

S

T

U

V

Z

FSC
www.fsc.org
MIX
Papier | Fördert gute Waldnutzung
FSC® C083411

Kolibri 360 GmbH
Ferdinand-Jühlke-Straße 7
99095 Erfurt, Deutschland
produktsicherheit@kolibri360.de